YOU'RE NOT MAD

UNDERSTANDING AND DEALING WITH NARCISSISM

DONNA SIGGERS

This book aims to help victims and survivors of narcissistic relationships gain clarity, rebuild self-worth, and regain emotional control. It serves as a guide for understanding and positive change.

This book is written and edited in British English rather than US English. This includes spelling, grammar, and punctuation.

"Instead of praising people for being 'resilient' change the systems that are making them vulnerable"
Muna Abdi

"If someone treats you badly, please remember that there is something wrong with them, not you. Normal people don't go around destroying other human beings"
The Narcissist Detox

CONTENTS

PART ONE

UNDERSTANDING ABUSE

"The effects of trauma can be devastating for sufferers, their families, and future generations"
Bessel Van Der Kolk

Before examining the complexities of narcissistic relationships, it is important to clarify some terminology. The term *narcissism* originates from Greek mythology: Narcissus, known for his beauty, rejected the nymph Echo, as well as others, after which he fell under the curse to fall in love with his own reflection. He subsequently spent his days fixated on his image. This narrative introduced two key concepts that persist today: an intense focus on oneself and a lack of consideration for others. The modern use of the term appeared in 1889 when Paul Nacke, a German psychiatrist, described narcissism (narzissmus, in German) as a sexual perversion, referring to individuals who regarded themselves as their primary sexual object. At that time, perversion included behaviours now considered normal, such as masturbation. Freud later proposed a connection between narcissism and male homosexuality (now known as same-gender relationships) by suggesting attraction to men involved self-directed desire, a theory that is now outdated.

A point of confusion often arises regarding the distinction between *self-confidence* and narcissism. People with healthy self-confidence can consider others' perspectives, exhibit empathy, and are not solely focused on themselves. By contrast, narcissists tend to concentrate inwardly, often lacking empathy and projecting confidence that can attract vulnerable people, sometimes leading to manipulation.

The term *co-dependent* is commonly used for those involved in relationships with narcissists; however, it may not be clinically precise. Often, an individual in this situation exhibits qualities such as compassion, loyalty, and adaptability, which can be subject to influence by narcissistic partners. Behaviours often described as co-dependency may instead reflect adaptive coping mechanisms or trauma responses rather than inherent personality traits. The use of these labels has the potential to obscure genuine characteristics. Emotional intelligence (EI) is a strength that often demonstrates an ability to understand multiple perspectives: Those displaying EI are dependable, capable, and maintain consistency within their relationships, which can make them susceptible to manipulation through guilt or shame. Oftentimes there is a higher tolerance for discomfort which can lead to persistence in challenging or unstable situations, reflecting resilience under stress. Loyalty and commitment are notable aspects of behaviour, which may be leveraged by narcissistic partners to sustain control. In addition, those open to personal growth or displaying ambition, competence, and self-confidence—attributes considered valuable—may encounter conflict due to a narcissist's attention-seeking and perceived need for superiority. Narcissistic people may assign responsibility to their partners in instances where they themselves do not accept it. Support strategies typically emphasise strengthening personal assets and developing a clear sense of identity, rather than focusing on possible shortcomings.

In healthy relationships, both parties provide mutual support, reassurance, and avenues for expressing concerns and ambitions—key elements of meaningful human connection. In contrast, co-dependency is defined by excessive reliance on another person or the relationship itself, to the extent that one's identity becomes deeply entwined with maintaining the bond at any cost. Such dynamics often result from underlying issues, such as addiction, abuse, or untreated mental health conditions, which foster maladaptive behaviours. In co-dependent relationships, one partner frequently neglects their own needs and boundaries in favour of prioritising the other's wellbeing, thereby impeding resolution of fundamental problems. This pattern leads to an imbalance where one party consistently makes disproportionate sacrifices while the other receives unreciprocated support, inhibiting both to confront and resolve dysfunctional patterns. Indicators of co-dependency include compulsive attention to another, coupled with a belief that functioning independently is not possible, resulting in a deficit of interrelationship emotional independence, including subordinance to their own perspectives; concealing personal opinions for approval; and experience a significant fear of abandonment, sometimes resulting in emotional suppression; dishonesty; or enabling problematic behaviours in their partner. Those experiencing co-dependency often lack external support systems, limiting social interactions due to an overwhelming focus on one individual at the expense of other relationships. A diminished sense of self, uncertainty regarding personal identity, and increased deference to others' judgment further characterise this state. Over time, co-dependent partners may develop feelings of resentment while simultaneously believing they cannot function without the other person, signalling an unhealthy relational dynamic.

Common traits of co-dependency encompass guilt for prioritising personal needs; accepting undue blame to avoid conflict; assuming excessive responsibility; lacking fulfilment outside the relationship; ignoring or excusing detrimental behaviours; and focusing solely on a partner's positive attributes. These patterns can erode self-esteem and hinder recognition of one's emotions, resulting in challenges with decision-making, overemphasis on caregiving, seeking validation through relationships, and heightened responsibility for others' actions. These factors may also increase the likelihood of forming attachments with people around them who require support. Additional hallmarks include difficulty adapting to change; pronounced need for approval; sensitivity to unmet expectations; and a tendency to attempt to manage others, all of which may contribute to deficiencies in communication skills. Understanding these distinctions is essential for comprehending the mindset of individuals with narcissistic traits and their impact on the wellbeing of those around them.

Narcissistic abuse is not confined to any specific community, profession, or economic status; instead, it can affect individuals from diverse backgrounds and with varying personality types. This book seeks to provide an in-depth examination of narcissism, its typical behaviours, methods for recognising narcissistic relationships, and approaches for effectively managing these interactions. The following statistics, recorded in the statistical year 2024 (April-March), from The Office For National Statistics (www.ons.gov.uk) indicate the prevalence—and seriousness—of this form of abuse:

An estimated 712,000 men in England and Wales experienced domestic abuse in the United Kingdom.

An estimated 1.6 million women experienced domestic abuse in the United Kingdom.

On average, one woman is killed by an abusive partner, or an ex-partner every five days in the United Kingdom.

Young girls in the United Kingdom reported high incidence of sexual violence—41% of girls aged 14-17 who are in intimate relationships are likely to experience some sort of sexual violence from their partner.

Domestic abuse prosecutions only totalled 51,183 in England and Wales during this period.

Additionally, according to UN Woman:

85,000 women and girls were victims of femicide, internationally during 2023.

51,100 of these killings were carried out by intimate partners or ex-partners—equating to 1,635 deaths per week.

"A stiff apology is a second insult—the injured party does not want to be compensated because he has been wronged; he wants to be healed because he has been hurt"
G K Chesterton

Individuals with narcissistic tendencies may find it challenging to demonstrate authentic remorse, making it essential to differentiate between *sorry* and *apology*. The term *sorry* typically conveys personal regret or sympathy and is often perceived as more heartfelt. In contrast, *I apologise* tends to be more formal and may lack sincerity. For instance, the phrase *sorry for your loss* communicates sorrow, emotion, empathy, and sincerity. Contrastingly, *I apologise for your loss* appears formal and less personal, conveying regret and formal responsibility. In relationships involving narcissistic traits, sincere apologies may be uncommon. Maintaining composure and establishing boundaries can support individuals in safeguarding their wellbeing. Those with narcissistic personality disorder, or characteristics, may not readily acknowledge mistakes, potentially due to a strong sense of entitlement.

Chapter One

NARCISSISTIC PERSONALITY DISORDER

*"Hate is the complement of fear and narcissists
like being feared. It imbues them with an
intoxicating sensation of omnipotence"*
Sam Vaknin

Narcissistic Personality Disorder (NPD) is a recognised mental health condition characterised by a pervasive pattern of grandiosity, a constant need for admiration, and a lack of empathy, which may result in maladaptive behaviours. NPD may manifest alongside attributes such as intelligence, charisma, or perceived success. Evidence suggests narcissistic personality disorder is diagnosed more frequently in males than in females. Narcissism can be overt, covert, antagonistic, communal, or malignant. When considering narcissism as a personality trait in relation to its impact on daily functioning and interpersonal relationships, these five types of narcissism can be organised into two primary categories: Adaptive (helpful) and maladaptive (unhelpful).

Narcissism involves increased self-focus, a heightened perception of self-importance, and a strong need for recognition. Understanding the traits and classifications of narcissism can provide insights into related cognitive processes, emotional responses, and behavioural tendencies:

Overt narcissism, also referred to as *grandiose or agentic narcissism*, is a commonly identified form of narcissistic personality. Individuals may display outgoing behaviour; assertiveness; entitlement; and an emphasis on their self-image, along with seeking admiration and exhibiting limited empathy. This subtype is linked with the 'big five' personality traits of extraversion and openness. Studies suggest that people with overt narcissism often report higher self-esteem and are less likely to disclose feelings such as sadness, worry, or loneliness. Furthermore, these individuals may assess their own abilities, intelligence quotient, and emotional intelligence more favourably than average.

Covert narcissism, also called *vulnerable or closet narcissism*, contrasts with overt narcissism. While narcissism is often seen as loud and overbearing, covert narcissists display more subtle traits such as expressions of low self-esteem; higher likelihood of experiencing anxiety, depression, and shame; introversion; insecurity or low confidence; defensiveness; avoidance; and the tendency to feel or play the victim. They often display a strong self-focus, which may coexist with concerns about personal adequacy. Research indicates that covert narcissism is correlated with higher neuroticism (a tendency to experience negative emotions) and lower agreeableness. Those with covert narcissism may find it difficult to accept criticism and, compared to individuals with overt narcissism, are more likely to interpret criticism more intensely than intended.

Antagonistic narcissism is a subtype of overt narcissism and is characterised by a focus on rivalry and competition. Key features of this subtype include arrogance, propensity to exploit others, competitive behaviour, and disagreeableness or an inclination toward argumentation. Individuals exhibiting antagonistic narcissism are typically less inclined to forgive others and may demonstrate reduced levels of trust in interpersonal relationships.

Communal narcissism is a form of overt narcissism, often described as contrasting with antagonistic narcissism. Individuals with communal narcissism typically place importance on fairness and may view themselves as altruistic, though there can sometimes be inconsistencies between these self-perceptions and their actual behaviours. People exhibiting communal narcissism might become easily morally outraged, describe themselves as empathetic and generous, and react strongly to things they see as unfair. Communal narcissism differs from genuine concern for others in its emphasis on social power and self-importance. For instance, those with communal narcissism may not recognise that their actions toward others contradict their self-image.

Malignant narcissism varies in severity and often impacts the individual more than others. It is more associated with overt than covert narcissism. Features of malignant narcissism might include vindictiveness, sadism or getting enjoyment from the pain of others, aggression when interacting with other people, and paranoia, or heightened worry about potential threats. Individuals exhibiting malignant narcissism may present with certain characteristics that overlap with Antisocial Personality Disorder (APD). As a result, those with malignant narcissism could have an increased risk of encountering legal issues or developing substance use disorders. Additionally, they often experience greater difficulty managing anxiety and enhancing their day-to-day functioning.

The DSM-5-TR lists nine criteria, of which five must be present, for diagnosing Narcissistic Personality Disorder:

A *grandiose sense of self-importance* is the practice of exaggerating achievements, talents, and expecting to be seen as superior without commensurate accomplishments.

Frequent Fantasies about having or deserving success, power, intelligence, beauty, love, and self-fulfilment.

A *belief in superiority*, believing they are special or unique, that they should only associate with those they see as worthy.

A *need for excessive admiration* due to fragile self-esteem, frequent self-doubt, self-criticism, or emptiness. A person would be preoccupied with knowing what others think of them and will fish for compliments.

Entitlement that displays as an inflated sense of self-worth, expectation of favourable treatment to an unreasonable degree, and anger when people do not cater to or appease them.

A *willingness to exploit others* through consciously or unconsciously using others, forming friendships or relationships with people who boost their self-esteem or status, and deliberately taking advantage of others for selfish reasons.

They *lack empathy*, saying things that might hurt others; seeing the feelings, needs, or desires of others as a sign of weakness; and not returning kindness or interest that others show.

Experience *frequent envy* that shows as feeling envious of others especially when others are successful, expecting envy from others, and belittling or diminishing the achievements of others.

Arrogance showing as patronising behaviour, behaving in a way that is snobby or disdainful, and talking down or acting condescendingly.

In some circumstances, the criteria associated with narcissistic personality disorder may include experiencing vulnerability, withdrawing from social interactions due to such feelings, exhibiting perfectionism regardless of concern for failure; heightened sensitivity to criticism, perceiving rejection or failure, displaying depressive symptoms following rejection or failure, reacting with anger or frustration when criticised or rejected, demonstrating humility as a strategy for self-image management, and avoiding situations where failure is possible, all of which may affect overall achievement.

The etiology of narcissistic personality disorder remains inconclusive. Research has demonstrated subtle distinctions in brain structure between individuals diagnosed with NPD and those without the disorder; however, it is not yet established whether these neurological differences are causes or effects of NPD. Current evidence indicates that both genetic predispositions and social learning processes may contribute to the development of this condition. For example, the presence of similar traits in close relatives and the modelling of behaviours by significant individuals during formative years likely contribute to this condition. Furthermore, certain parenting styles—such as overindulgence and overprotection—may lead children to expect continual preferential treatment, potentially impeding the development of emotional regulation skills, particularly during times of adversity. Additionally, there is an observed correlation between adverse childhood experiences, including trauma, rejection, neglect, or insufficient support, and the emergence of narcissistic traits.

Complications often arise as overlapping features between narcissistic personality disorder and other mental health conditions. For example, mood disorders and bipolar disorder are associated with an increased risk of anxiety and depression, with bipolar disorder being observed more frequently in individuals with NPD. Narcissism may co-occur with Borderline Personality Disorder (BPD) or Antisocial Personality Disorder (ASPD). Additionally, Body Dysmorphic Disorder may be present when negative perceptions about personal appearance increase susceptibility. Some individuals diagnosed with NPD may use alcohol or substances, especially stimulants such as cocaine, as a coping mechanism, which can contribute to the development of Substance Use Disorder (SUD). People with NPD may experience significant depressive symptoms when faced with setbacks, failure, or rejection. In these instances, suicidal behaviour is less impulsive and has a higher likelihood of resulting in completed suicide.

A psychiatrist or psychologist assesses whether an individual has NPD by collecting information about their life and relationships, and by asking further questions to distinguish it from similar conditions. Treatment options include therapeutic approaches such as Dialectical Behavioural Therapy (DBT), Cognitive Behavioural Therapy (CBT), Metacognitive therapy, group therapy, and couples or family therapy. Medications, including antidepressants, antianxiety medications (anxiolytics), and antipsychotic drugs, may be used to address related issues like anxiety and depression. Currently, there are no medications specifically approved for NPD.

Individuals diagnosed with narcissistic personality disorder frequently lack awareness of their condition. Symptoms may become apparent during periods of stress or adverse circumstances, such as job loss, demotion, bankruptcy, personal crises like breakups or divorce, age-related illnesses, reduced ability to live independently, or bereavement. Those with NPD typically do not seek treatment voluntarily, often only considering intervention at the urging of family members or others.

The prognosis for narcissistic personality disorder depends on the specific symptoms exhibited. Individuals displaying vulnerability-related characteristics may be less inclined to seek assistance or disclose their condition. Conversely, when symptoms include pronounced aggression or reduced empathy, daily functioning is often more significantly impaired and treatment engagement tends to be lower, thereby further complicating outcomes. These conditions include anxiety, depression and other mental health illness such as bipolar disorder, antisocial personality disorder, and substance abuse disorder. Additionally, individuals diagnosed with narcissistic personality disorder may not accept their diagnosis, which can result in resistance to treatment.

Chapter Two

NARCISSISM AS A PERSONALITY TRAIT VERSUS PERSONALITY DISORDER

"If you hand them the strings, they become the narrator of your life"

Narcissism can be viewed as existing along a spectrum, with some individuals exhibiting narcissistic traits without necessarily meeting criteria for a personality disorder. Such traits may be mild or appear in specific contexts but could also be present in a range of situations. If these behaviours are prevalent and persist over time, they may lead to a diagnosis of narcissistic personality disorder by a mental health professional. Individuals with narcissistic traits may sometimes recognise their impact on others, whereas those diagnosed with NPD often experience greater difficulty in acknowledging responsibility. A diagnosis of NPD requires symptoms to manifest in all aspects of life, while narcissistic traits may only be observed in limited circumstances. For example, a person may display strong ambition at work yet remain receptive to feedback in personal relationships.

Chapter Three

AVOIDING RESPONSIBILITY

Narcissist's Prayer

That didn't happen
And if I did, it wasn't that bad
And if it was, that's not a big deal
And if it is, that's not my fault
And if it was, I didn't mean it
And if I did...you deserved it

A tendency to avoid responsibility for ineffective or problematic behaviour is sometimes seen in individuals exhibiting narcissistic personality traits or NPD. These individuals may demonstrate outward confidence that conceals underlying low self-esteem. As part of their coping strategies, they infrequently acknowledge mistakes, except on trivial matters which can serve multiple functions. The reluctance to recognise wrongdoing or accept accountability for adverse actions frequently involves denial—such as denying events, other people's actions, consequences, or aspects of reality. Over time, persistent denial may become habitual and affect their perception of reality. In certain instances, a phenomenon known as *toxic amnesia* is observed, wherein an individual asserts a lack of recollection regarding harmful actions, betrayal, dishonesty, or related behaviours. This conduct has been identified as a manifestation of gaslighting, aimed at undermining another person's perceptions or memories. Additional tactics may include attributing fault to others or adopting a victimised stance. Such strategies function to shift responsibility away from themselves. These patterns can occur not only within interpersonal relationships but also on a societal scale; for example, certain propaganda efforts may depict adversaries as both formidable and vulnerable, depending on narrative objectives.

The prayer at the beginning of this section explores themes related to narcissism. Typical response expectations from the narcissist are likely to be unrealistic and self-justifying towards their behaviour. For example:

That didn't happen (pure denial, toxic amnesia, gaslighting) with the expected response, *You're right, I misunderstood something. I'm sorry.*

And if it did, it wasn't that bad (denial, minimisation) with the expected response, *You're right. It wasn't that bad, I overreacted. Sorry.*

And if it was, that's not a big deal (denial, minimisation) with the expected response, *You're right. I'm sorry, it's nothing. I shouldn't have brought it up.*

And if it is, that's not my fault (denial, rejection of responsibility, deflection) with the expected response, *You're right, I really overreacted. It's not your fault.*

And if it was, I didn't mean it (denial, lying, rejection of responsibility) with the expected response, *I know you wouldn't hurt me. It's okay.*

And if I did…you deserved it (denial, blaming the victim, deflection) with the expected response *I'm sorry, I didn't mean to act this way. It's all my fault, I'm very sorry.*

Individuals exhibiting narcissistic traits may seek to maintain their self-esteem by redirecting responsibility for behaviours that could be viewed as maladaptive. Common strategies include denial, cognitive distortion, selective memory, manipulation of perceptions (gaslighting), minimisation, deflection, externalising blame, and adopting a victim posture to bolster their self-image.

Chapter Four

DARK PSYCHOLOGY MANIPULATION TACTICS

---

*"Not all monsters hide in the shadows; some
smile in daylight, too"*

Dark psychology examines the use of psychological knowledge about human behaviour and emotions to influence others, including techniques such as guilt induction, gaslighting, and reverse psychology. Some individuals demonstrate emotional intelligence (EI), which enables them to empathise, adapt to various social environments, and resolve conflicts. These abilities can also be used for manipulative purposes. Dark psychology explores how traits such as Machiavellianism (a personality trait construct characterised by manipulativeness, indifference to morality, lack of empathy, and calculated focus on self-interest); psychopathy; and narcissism relate to the understanding and application of emotions and social interactions to affect other's actions.

The term dark psychology refers to the use of psychological principles with the aim of influencing, manipulating, or controlling others, which may result in negative outcomes. It includes tactics such as manipulation, coercion, and deception used for personal purposes. While dark psychology is not recognised as a formal scientific field, it is a term often used in general discussion to describe these practices. Common methods may involve manipulation, such as gaslighting, which can affect another individual's actions, thoughts, or decisions, along with approaches like persuasion, deception, and coercion. Deception may also take the form of sharing accurate information while omitting key details, known as paltering. The intent behind employing dark psychology techniques typically involves causing someone to question their perceptions or cognitive understanding through denial of facts or altering reality. In interpersonal relationships, this may include challenging or invalidating a partner's memories or experiences, potentially leading to self-doubt. Gaslighting and other forms of manipulation can coexist with additional adverse behaviours and may gradually diminish an individual's self-confidence and trust in others. Additionally, opposing another's interests without their knowledge after building trust may serve to benefit one party at the expense of another, an action sometimes referred to as backstabbing. Narcissists may involve third parties (triangulation) to influence or control situations, often by making negative comments designed to provoke feelings of jealousy or insecurity, which can undermine new relationships. They may exploit an individual's emotional, psychological, or physical vulnerabilities to manipulate them, targeting perceived weaknesses. For example, a scammer might form a relationship with an isolated older adult to persuade them to transfer financial assets. Another tactic is guilt tripping, wherein an individual uses guilt to influence another's actions by making them feel unduly responsible for outcomes beyond their control or exaggerating the impact of their behaviour. Reverse psychology is also frequently employed, whereby suggestions are made with the intent of eliciting the opposite response. For instance, an individual may encourage their partner to spend time with friends while simultaneously expressing a preference for their company instead of remaining at home with them. Other tactics include love bombing and negging. Love bombing refers to a manipulative strategy in which an individual inundates another person with excessive affection, compliments, and gifts to rapidly establish trust and emotional attachment, typically as a precursor to exerting control within romantic relationships. Negging is also a manipulative approach used in interpersonal contexts, intended to undermine an individual's confidence and prompt them to seek validation from the initiator. This behaviour typically includes backhanded compliments or subtle insults.

Individuals exhibiting characteristics of Machiavellianism, psychopathy, or narcissism are statistically more inclined to employ dark psychology strategies and are conducted by abusers, criminals, people with personality disorders, and those with neurotic traits. Those scoring highly on measures of neuroticism typically experience increased emotional distress, anxiety, and mood fluctuations. Research suggests that someone with these personality traits is more inclined to employ tactics like guilt induction or threats to end relationships, while being less likely to resort to begging, compared to those with lower levels of neuroticism. They also tend to use manipulation strategies more frequently in daily interactions. Resulting behaviours of victims associated with neuroticism may include preferring long, hot showers due to a heightened desire for physical and mental comfort, consistently selecting window seats out of a need for control and personal space, and finding limited enjoyment in social situations as their cognitive processing may filter casual conversation as irrelevant. Additionally, self-talk is not indicative of irrationality; rather, it reflects heightened cognitive functioning and supports effective decision-making.

Chapter Five

SIGNS, CAUSES, AND TYPES OF MANIPULATIVE BEHAVIOURS

"The most dangerous thing about narcissists is their ability to control and manipulate in order to get what they want"

Narcissistic individuals frequently employ manipulation as a means of exerting control over situations and others. This conduct involves crossing personal boundaries and is intended to secure power within relationships. Identifying common indicators of such behaviour can assist in safeguarding against these methods, which may become more subtle when narcissists experience rejection or diminished control. Although manipulation can be difficult to detect due to its subtlety, it often leaves affected individuals with a sense of discomfort. Manipulation can be pro-social (beneficial to relationships), or anti-social (harmful to relationships).

Accurately identifying manipulative behaviour requires careful consideration of an individual's developmental stage. For example, young children may display temper tantrums due to limited verbal abilities to express their emotions; such conduct is developmentally expected, as most children gradually acquire more effective social critical thinking skills. In adolescence, manipulation can emerge in socially adaptive contexts—such as employing trivial lies to protect a peer's dignity or maintain group cohesion. By adulthood, individuals typically transition from overtly manipulative behaviours towards more constructive strategies, including practising assertiveness and setting healthy boundaries. Notably, those with histories of relational disruption or trauma may be more prone to engage in manipulation, aggression, or deceit as mechanisms for meeting their needs. These actions often reflect attempts to preserve power or control within relationships, sometimes despite an awareness of the potential harm inflicted on others. Recognising manipulative behaviour remains complex, as it is frequently disguised as well-intentioned conduct. Signs of a manipulative person include persistent and excessive attention, love, and flattery that cross comfortable boundaries; time pressures for others to complete tasks or act; incongruence between words and actions; and making the victim feel guilt, shame, or seem off around this person. There are many forms of influence found within manipulative relationships:

Coercive control, resulting in loss of identity, self-assurance, wellbeing, and sense of reality; the narcissist changing the rules and/or standards for achieving the finish line.

Emotional blackmail that appeals to address needs, including the potential disclosure of sensitive or personal information; gaslighting that challenges or minimises the validity of reported observations, experiences, or emotions.

Guilt tripping, that creates a sense of obligation or indebtedness in recipients through perceived acts of generosity; isolation that restricts a victim's contact with their support network or community

Deception through disclosing selective aspects of the truth, providing fabricated information; love bombing, providing praise, gifts, and attention to individuals.

Passive aggression, wherein the narcissist communicates dissatisfaction in an indirect manner, such as through back-handed compliments.

Projection, or attributing responsibility for certain actions to the victim, even though those actions are being performed by another individual.

Silent treatment during which the victim's efforts to initiate communication are disregarded.

Smear campaigns, during which negative remarks or disseminating rumours about an individual without their awareness occurs.

Triangulation, where a third party becomes involved in communication to increase influence; and triggering insecurities by making remarks or comments presented as jokes that may lower confidence.

Persistent manipulative behaviour can result from multiple factors and often begins in childhood. Such tendencies may develop when direct needs are unmet or in environments involving conflict, competition, or abandonment, and these patterns may continue into adulthood. While some mental health conditions and interpersonal styles are associated with manipulation, manipulative behaviours can also arise independently of these factors. Manipulation is defined as influencing another person's thoughts or feelings through psychological means, which may cause individuals to reconsider their perceptions, feel responsibility for specific situations, or experience uncertainty. Manipulative mental health conditions include antisocial personality disorder, borderline personality disorder, co-dependency, historic personality disorder, insecure attachment style, narcissistic personality disorder, and substance use disorder. No personality type is inherently manipulative but when traits are present, it is considered a behaviour that can occur in varying degrees, and can become persistent, occurring across multiple situations,

Those subjected to emotional manipulation frequently experience pervasive feelings of guilt and a range of adverse outcomes, including anxiety, depression, heightened stress levels, diminished life satisfaction, reduced self-esteem, and increased social withdrawal. These effects often lead individuals to question their own perceptions and decisions, resulting in chronic uncertainty and challenges in trusting both themselves and others.

Chapter Six

SIGNS OF NARCISSISTIC RAGE

"Love doesn't die a natural death. Love has to be killed, either by neglect or narcissism"
Frank Salvato

Narcissist rage is a term introduced by author, Heinz Kohut, in 1972. It refers to responses observed in individuals with narcissistic personality disorder, who may display strong anger following what may seem like minor or ambiguous provocation. In situations where these individuals do not receive continual positive feedback from others, this can result in feelings of shame that prompt an immediate angry reaction, sometimes expressed without regard for its effects on others. This behaviour has been linked to sensitivity to perceived criticism and concerns about maintaining self-image. Narcissistic rage is characterised by a pronounced and intense anger response that arises in reaction to perceived threats to self-esteem. In contrast to typical expressions of anger, this reaction is frequently disproportionate, unpredictable, and may involve verbal or physical aggression, manipulative behaviours, or passive-aggressive actions. Such episodes can manifest with both internal and external symptoms, often catching affected individuals by surprise. Rage carries with it outward and inward signs. Outward signs include episodes of anger when the desired attention is not received, a raised voice or consistent shouting, a sudden onset of angry emotions that are conveyed to victims, difficulty managing anger, and deliberately directing emotional or physical aggression towards others. Inward signs include indirect expression of emotions, withdrawing or demonstrating aloofness, avoiding interactions with others, experiencing feelings of resentment, failing to complete tasks, using sarcasm in communication, displaying strong feelings of indignation, exhibiting a sense of entitlement, showing hostility or bitterness, disengaging from people to safeguard self-esteem, and experiencing dissociation or a sense of disconnection from reality.

The precise causes of narcissistic personality disorder, which is frequently linked to episodes of narcissistic rage, have yet to be fully established. Current research indicates that genetic factors, early developmental experiences, and significant life events may all play a role in the onset of NPD. Assessing whether an individual exhibits traits of NPD, or related narcissistic features, can understandably present challenges. Several contributing elements are recognised within the context of both NPD and narcissistic rage. Furthermore, individuals manifesting narcissistic rage as a reaction to perceived threats may be considered to have experienced a narcissistic injury. There are many factors to be considered:

Early childhood trauma: Experiences of emotional abuse or neglect may lead individuals to suppress their authentic self and develop alternate personas as coping mechanisms.

A highly sensitive temperament: Individuals with heightened sensitivity to feelings such as shame might display intensified emotional responses under stress.

Failure to develop critical emotion regulation skills: Insufficient development of emotion regulation skills may result in less adaptive reactions to situations.

An unstable sense of self-esteem: Instability in self-esteem can contribute to anxiety about personal vulnerabilities being exposed and may influence behavioural responses.

Facing a setback or disappointment: Setbacks and disappointments may elicit strong emotional reactions, particularly in those who closely tie self-image to external outcomes.

Envy: Feelings of envy towards other's possessions, relationships, or status may sometimes be associated with increased emotional reactivity.

Memories of early experiences of shame: Current events can trigger memories of early shame, potentially resulting in intense emotional responses.

Splitting: Perceiving situations in extremes (e.g. all good or all bad), may relate to shifts in attitudes toward people and can influence emotional reactions.

Having a split sense of self: A perception of self-distinct aspects may affect an individual's ability to regulate emotions effectively.

A fragmented sense of self: A sense of self that is heavily influenced by external validation can create a reliance on other's perceptions rather than an internally derived self-concept.

Narcissistic rage can be categorised into two forms: *explosive rage* and *passive rage*. Individuals exhibiting explosive rage may use verbal insults, raise their voices, make threats, or display signs of self-harm. Those experiencing passive rage may withdraw and become unresponsive or disengaged for a period of time. Narcissistic rage differs from typical anger in that it often unfolds through distinct stages. Initially, *stress* may manifest as subconscious feelings of anger without any overt expression by the individual. This can lead to *anxiety*, marked by subtle signs indicating the onset of anger. The person may then become agitated, displaying visible dissatisfaction without explicitly assigning blame. The *irritation* phase is characterised by verbal cues intended to prompt a response or change from others. Frustration may emerge through changes in facial expression or the use of emphatic language. As *anger* intensifies, there is typically an increase in vocal volume and expressiveness. Rage can result in a loss of composure and episodes of aggressive behaviour.

In contrast to typical responses, a narcissistic reaction often escalates from stress to rage, with anger directed either outwardly or inwardly. Such experiences can have various effects on both the individual and those around them. Rage may impact several aspects of life, such as family relationships, social connections, academic or professional achievements, finances, job or school attendance, legal matters, and health. It can also manifest as violence toward others or self-injury, and may contribute to feelings like guilt, loss, difficulty adapting to change, depression, anxiety, physical health concerns, substance use, and suicidal thoughts or behaviours. Individuals experiencing narcissistic rage might lack awareness of their emotional state; while these reactions may temporarily reduce feelings of fear and shame, they often lead to social isolation and affect wellbeing. Therapeutic support can help identify behaviours, manage internal conflicts, address underlying issues, and develop more adaptive coping strategies. Therapists can assist individuals with narcissistic traits in gaining self-awareness and exploring their sense of identity; support the evaluation of whether maintaining current behaviours outweighs the benefits of making changes; facilitate the development of a more resilient self-concept, independent of external validation; provide help in processing past traumatic memories or experiences associated with feelings of shame that may arise when anger becomes problematic; offer support in adapting to life without relying on previous coping mechanisms such as self-enhancement and manipulation; clarify the role of fear of rejection in anger responses and how this may contribute to ongoing cycles of rejection; promote learning and practicing healthy relationships, both with oneself and others; guide individuals through the process of addressing feelings of inadequacy and a fragile self-image; and foster individual empowerment.

Chapter Seven

POWER STRUGGLES IN RELATIONSHIPS

"The balance of power is the scale of peace"
Thomas Paine

Power dynamics are prevalent within human relationships, manifesting in both personal and professional contexts. Such dynamics frequently arise from divergent needs, desires, or perspectives among individuals, resulting in conflict and strained interactions. These circumstances can be emotionally taxing and may impede individual growth, relationship development, and overall wellbeing. Difficulties in achieving consensus often lead to challenges in compromising or making joint decisions on significant matters, which may indicate an imbalance of power or insufficient mutual respect. Persistent patterns of power struggles tend to foster resentment and gradually undermine the foundational aspects of relationships.

When individuals perceive that their opinions or needs are not acknowledged in discussions—such as one party dominating the conversations, consistently interrupting, or dismissing concerns—this points to underlying power struggles. The repeated emergence of unresolved disagreements without substantive progress further signals deeper issues related to power imbalances. Over time, such dynamics can lead to the accumulation of resentment and hinder open communication. A pronounced indicator of a power struggle is when one individual consistently feels marginalised or disregarded during decision-making processes, including unilateral decision-making or control over resources and opportunities.

It is essential to recognise that power is not inherently negative; it can facilitate collective achievements and drive change. Problems emerge when power imbalances result in feelings of diminishment or exclusion. Often, such struggles stem from fundamental differences in values or fears related to autonomy and individuality, fostering resistance to compromise or efforts to maintain control. Cultural norms and societal expectations also play a role in shaping assumptions about roles and responsibilities within relationships. For example, traditional gender roles might implicitly assign financial management or household duties, potentially leading to friction if these expectations are not explicitly addressed and mutually agreed upon.

Power struggles commonly occur when individuals seek to assert control or resist perceived dominance by others, presenting in various forms such as disputes over decision-making, competition for recognition, or disagreements regarding boundaries and responsibilities. At their core, these conflicts reflect deeper emotional needs and insecurities, and when left unaddressed, they erode trust and contribute to ongoing resentment. The nature of common power struggles varies depending on the context. In personal relationships, such issues often revolve around decision-making authority and independence. Parent-child dynamics typically involve conflicts over autonomy and boundary-setting, whereas workplace power struggles may relate to role clarity, advancement opportunities, and leadership authority. Even friendships can experience strain due to competing interests and influence. While disagreements are unavoidable in any relationship, adopting a respectful approach is crucial for maintaining healthy connections. Addressing underlying concerns, enhancing communication skills, fostering empathy, and developing effective conflict resolution strategies can significantly mitigate the adverse effects of power struggles and promote more constructive interactions.

Chapter Eight

POWER IMBALANCE WITHIN ABUSIVE RELATIONSHIPS

"Peace is not the absence of conflict but the presence of creative alternatives for responding to conflict—alternatives to passive or aggressive responses, alternatives to violence" Dorothy Thompson

While power dynamics or imbalances in relationships are frequently seen as undesirable, they do not inherently result in negative consequences. Early in a relationship, partners may find fulfilment in valuing each other's unique perspectives, beliefs, and viewpoints. In certain instances, power imbalances can foster personal growth and enhance mutual understanding and respect. Although challenges may arise, recognising the individual strengths each person contributes can facilitate the development of equilibrium within the relationship. Such shared awareness is instrumental in clarifying boundaries and promoting constructive compromise.

Abusive relationships frequently involve dynamics of fear, manipulation, and control that can substantially hinder an individual's capacity to disengage. Many individuals encounter significant obstacles when attempting to leave such circumstances. A comprehensive understanding of the seven stages of trauma bonding offers critical insights into these challenges and may guide effective intervention strategies. Trauma bonding is typically marked by loyalty, dependency and, occasionally, affection, despite ongoing negative behaviours.

The origins of power dynamics in relationships can be understood within two primary frameworks: *unhealthy* and *healthy power dynamics*. Unhealthy power dynamics involve the capacity to influence or modify another individual's behaviour, independent of explicit dominance or submission, whereas healthy power dynamics reflect a balanced interplay in which each persons' ability to influence and contribute is acknowledged and nurtures constructive growth within the relationship. The possession of power significantly impacts psychological processes, often without conscious awareness. Empirical studies suggest that power activates the behavioural approach system in the left frontal cortex, resulting in heightened dopamine activity through automatic mechanisms. Furthermore, occupying a position of power tends to influence behaviour and decision-making, frequently shifting focus toward goal achievement rather than interpersonal relations. Collectively, these factors help explain recurring patterns observed in settings characterised by power imbalances.

In some partnerships, one individual may possess a greater degree of authority and influence, often taking the lead in key decision-making processes. This arrangement can result in that person serving as the primary decision-maker, with their partner typically deferring to their judgment. In situations of disagreement, the same individual is more likely to assert their preferences, while the other partner tends to acquiesce. The dominant partner frequently initiates conversations, structures discussions, and guides outcomes, sometimes without the explicit awareness of the other party. Collaborative deliberation may occur, yet it is not always recognised by the less influential partner until later, when they realise that they were persuaded.

Examining early developmental stages can provide valuable insights into why certain individuals may exhibit narcissistic traits or become involved with those who do. In infancy, it is typical for children to view the world from an egocentric standpoint. When consistently exposed to positive feedback regarding their qualities, they often internalise these perceptions as reality due to limited contrasting experiences. Subsequently, if negative interactions occur later, such experiences may feel unexpected and distressing. Conversely, children subjected to abuse or raised in environments where mistreatment is prevalent may develop different expectations of what is considered normal or acceptable. Such formative experiences shape their self-concept and influence how they respond to adverse treatment in adulthood, potentially reducing their likelihood of resistance. These individuals might also have received occasional praise or positive reinforcement. Caregiver inconsistency, characterised by alternating supportive and negative behaviours, can foster trauma bonds. Furthermore, early life experiences and behavioural responses are frequently transmitted intergenerationally through learnt patterns, which may contribute to the manifestation of narcissistic traits within families.

Chapter Nine

UNHEALTHY POWER DYNAMICS

"Living an unbalanced life means understanding that on any given day, week, or year, every 'yes' we utter, means a no to countless other tasks and goals. It means embracing ebb and flow and the delightful truth that building a joyful life does not require perfection"
Krista O'Reilly-Davi-Digui

Unhealthy power dynamics can negatively impact both relational functioning and psychological wellbeing. When examining such dynamics within relationships, it is essential to establish a clear definition of power in this context. Power refers to the capacity to direct or influence another individual's behaviour in specific manners. Subsequent paragraphs elucidate key patterns of influential behaviours.

The demand/withdrawal dynamic is characterised by one partner persistently initiating discussion and seeking change regarding relationship concerns, while the other consistently withdraws and avoids engagement on these topics. Empirical research indicates that this pattern is associated with increased rates of spousal depression and serves as a significant predictor of divorce and marital dissatisfaction.

The distancer/pursuer dynamic emerges when one partner strives to achieve and sustain a certain degree of intimacy, whereas the other perceives such efforts as overwhelming. As the initiator intensifies their pursuit, the other party frequently responds with heightened resistance, defiance, and withdrawal. This interaction typically reflects differing needs for closeness rather than a struggle for dominance. The distancing partner may attribute relational issues to perceived dependency, while the pursuer may interpret the other's conduct as emotionally detached or unresponsive.

The fear/shame dynamic often operates at an underlying or unconscious level, contributing to relational challenges. In this dynamic, one partner's experience of fear and insecurity may elicit feelings of shame and avoidance in the other, and vice versa. These emotional responses are shaped by a range of factors, including hormonal influences and previous traumatic experiences.

Chapter Ten

HOW TO SPOT UNHEALTHY POWER BALANCE

"Balance is a feeling derived from being whole and complete; it is a sense of harmony. It is essential to maintaining quality in life and work"
Joshua Osenga

Power dynamics in relationships can influence interactions between partners. Examining specific aspects of power may help identify imbalances, whether positive or negative, in how power is shared. Achieving a balanced distribution of power may affect how couples communicate and make decisions together. In this arrangement, both individuals are responsible for themselves and for maintaining the relationship. Decisions are made jointly, and all perspectives are considered. Open communication is recommended when issues arise, allowing room for honest discussion and understanding. Key elements of shared power include attention, where the emotional needs of both parties are identified and addressed; influence, in which each partner can engage with and impact the other on an emotional level; accommodation, for situations such as hardship when prioritising one partner may be required, otherwise defaulting to joint decision-making; respect, demonstrated by consistent regard and appreciation for each other's humanity; selfhood, allowing each partner to maintain and value their individual identity within and outside the relationship; vulnerability, where both partners communicate their limitations and uncertainties transparently; and fairness, defined as a division of responsibilities and commitments that both parties consider equitable and supportive.

A fundamental component of sustaining a successful relationship is the ongoing evaluation of power dynamics. As both individuals evolve and confront new life challenges together, relational structures inevitably shift. It is essential to recognise that the distribution of power within relationships remains fluid and requires diligent attention. Negative imbalances of power are commonly characterised in three ways: demand-withdrawal, distancer-pursuer, and fear-shame patterns.

PART TWO

CHILD DEVELOPMENT

"Children are like wet cement. Whatever falls on them makes an impression"
Haim Ginott

Throughout life, interpersonal relationships play a significant role in shaping behaviour, often exerting both positive and negative influences. Lifelong learning involves the continual acquisition of information, comprising various perspectives, ideas, and beliefs introduced by others. From infancy, individuals acquire behavioural patterns through imitation and observation, with sensory experiences facilitating psychological development. These influences are particularly beneficial when they are constructive, supportive, and encouraging. As children mature, they gradually learn essential skills such as self-feeding, oral hygiene, dressing, forming friendships, riding bicycles, and caring for pets. Mastery of these abilities typically results from repeated practice and guidance, shaped by primary caregivers and other influential figures. A child's developing brain is not inherently equipped to differentiate between positive and negative behaviours; thus, children may imitate language or actions demonstrated by caregivers without discernment. Neurological development continues into early adulthood, with the frontal cortex—the last region to mature—reaching full development around age twenty-five. The frontal lobe, a component of the cerebral cortex, consists of paired left and right regions situated at the anterior portion of the skull near the forehead, an area often instinctively touched during periods of stress or confusion. Higher-level brain functions encompass voluntary movement control on the body's opposite side; sequencing of complex tasks such as dressing or preparing a hot beverage; facilitation of speech and language production within the dominant frontal lobe (typically opposite to hand dominance); regulation of attention and concentration; management of working memory for processing new information; support for reasoning and judgment abilities; organisation and planning skills, modulation of emotions and mood including interpretation of others' emotional states; expression of personality traits; motivation through evaluation of rewards and pleasure, regulation of impulse control; and oversight of socially appropriate behaviours.

Research indicates that positive encouragement and support from an early age are fundamental to healthy child development. A stable and calm environment allows children to learn through experience—making mistakes and learning from them—without undue chastisement. Ideally, this nurtures higher confidence and stronger self-esteem, enabling children to cultivate fulfilling relationships and lead well-adjusted lives. Exceptions do occur: Not all children will thrive in a nurturing environment.

The capacity of the human brain to focus is limited to [approximately] ninety to one-hundred-and-twenty minutes before requiring a break, which may explain attention shifts following extended periods of play or study. Interventions such as power naps, activity changes, meditation, and mindfulness can facilitate relaxation between intensive mental efforts. Higher cognitive functions—such as problem-solving, personality expression, motivation (including reward appraisal and pleasure), reasoning, and impulse control—develop at later stages. The wellbeing and future development of children are thus significantly influenced by their caregivers. If caregivers themselves are emotionally immature, have unresolved trauma, or were negatively shaped during their own upbringing, patterns of maladaptive behaviour may perpetuate across generations. Research has documented that unresolved trauma in caregivers can give rise to generational or transgenerational trauma. Transgenerational (or intergenerational) trauma refers to trauma originating decades prior to the current generation, shaping how individuals process and recover from adverse experiences. This phenomenon was first identified among descendants of Holocaust survivors, where increased psychiatric referrals were observed even among grandchildren, with rates up to 300% above those in the general population. Since then, intergenerational trauma has been identified in descendants of enslaved peoples, Indigenous groups, war survivors, refugees, and survivors of interpersonal violence, among others.

As individuals mature, many behaviours become ingrained, resulting in automatic responses informed by previous experiences— a process described as bottom-up processing. Children raised in earlier eras, particularly those now in their fifties and older, often grew up when children's rights were less respected; prevailing attitudes dictated that children should be seen but not heard, with parents retaining decision-making authority. Contemporary caregivers strive to impart appropriate values for today's society. However, automatic behaviours learnt unconsciously from previous generations may persist, especially if passed down from ancestors affected by trauma. Parents and caregivers frequently overrule children's choices, often with benevolent intentions and sometimes without awareness. Such actions can inadvertently overlook significant issues a child may wish to communicate but lacks the language or means to express. During these moments of automatic caregiving, early warning signs may be missed.

Chapter Eleven

TRAUMATIC CHILDHOOD EXPERIENCES

"Behaviour is the language of trauma. Children will show you before they tell you that they are in distress"
Micere Keels

A significant number of children face trauma before turning sixteen. This can include experiences such as sexual abuse, exposure to community violence, caregivers with substance use disorders, poverty, witnessing or experiencing domestic violence, bullying, bereavement, or serious illness. The impact of abuse does not necessarily disappear with age. Unresolved childhood trauma may lead to ongoing emotional and physical health issues in adulthood. Childhood trauma refers to distressing experiences that happen beyond a child's control. All family situations have the potential to become dysfunctional, creating opportunities for children to suffer traumatic experiences. As adults, individuals may opt to address these prior events to support their wellbeing. Identifying effective methods for managing the impacts of childhood trauma is recommended. Consulting professional mental health practitioners can assist in the healing process. Such professionals may help individuals develop healthier relationships, find value in their work, concentrate on physical wellness, and engage with others who share similar experiences. It is important to understand that trauma is the emotional reaction to frightening, extremely stressful, or life-threatening events (even if these incidents are perceived) and that people experience trauma differently, some coping well. Those who are struggling might experience depression, fear, anxiety, difficult memories, difficulty trusting others, feeling numb, and feeling detached.

In family systems involving children, the presence of at least one primary caregiver is essential. This role may be fulfilled by a biological parent, adoptive parent, foster carer, grandparent, or another member of the extended family. Additionally, households may include siblings, stepsiblings, adopted siblings, foster siblings, or other children related to the family. Ideally, family members maintain close relationships and interact with one another in both physically and emotionally healthy ways. It is important to recognise that familial roles are dynamic and may shift over time; children may assume different roles with each caregiver depending on the specific family dynamics. For example, a child could act as a 'hero' for one parent while functioning as a lost child with another. Significant changes within the family, such as divorce, can necessitate adaptation to new circumstances. Adult survivors of childhood trauma may identify with or relate to many of the following familial roles.

Within family dynamics, the caregiver's role typically involves providing care and support, while the child's role is to receive guidance and opportunities for development. Some caregivers may depend on their children in ways that are not age-appropriate, leading to changes in these roles.

> The term *parentified child* describes circumstances in which a minor is routinely expected to provide emotional or practical support to a caregiver, rather than receiving guidance and care themselves. This reversal of roles can influence the child's developmental trajectory and has been linked to long-term physical and mental health consequences. Parentified children may undertake responsibilities such as supervising siblings, maintaining household duties, or managing financial matters. Additionally, they may serve as confidants, offer advice, or

provide emotional reassurance. Studies indicate that ongoing parentification can affect neurological development, notably the hippocampus, which is integral to memory, emotional regulation, and stress management. In adulthood, individuals who experienced parentification during childhood may face challenges related to boundary-setting, emotional regulation, relationship formation, and may develop trust issues if caregivers were unreliable during formative years.

The *peacemaker child*, also known as the *hero child*, is frequently assigned significant responsibilities within the family unit, often managing various tasks and supporting other family members. As a result, these individuals tend to become high achievers who actively seek approval from caregivers. This role commonly emerges in families characterised by elevated expectations or where at least one caregiver is emotionally unavailable, prompting the child to pursue validation through accomplishment. Such children may develop an acute sense of accountability for family stability and assume mature responsibilities at a young age. While this experience can enhance their competence, it may also lead to increased pressure and perfectionistic tendencies. In adulthood, particularly within romantic relationships, former peacemaker children may struggle with vulnerability and link their self-worth to measurable achievements. This behavioural pattern sometimes results in prioritising a partner's needs over their own, contributing to potential relational imbalances. In friendships, they are often relied upon as sources of support, which may come at the expense of their own emotional wellbeing. Professionally, they typically exhibit strong performance but face an elevated risk of burnout due to persistent efforts to achieve success.

The *Scapegoat child* is typically designated as responsible for problems or negative outcomes within the family. This pattern can emerge in families where a parent often criticises, potentially leading the child to exhibit behaviours that reflect underlying issues. Such actions may draw attention away from broader family conflicts and contribute to behavioural challenges. As adults, individuals who have experienced this dynamic may face difficulties in romantic relationships, sometimes entering partnerships that mirror earlier experiences related to self-worth. They might display patterns such as engaging in specific relationship roles or encountering challenges with trust. In friendships, these individuals may oscillate between seeing themselves as sources of difficulty and seeking validation from peers. In professional settings, they may encounter conflicts with authority figures and experience feelings of misunderstanding or perceived unfair treatment.

The *invisible* or *lost child* is typically less prominent within the family system. This individual may receive limited attention due to perceptions that their needs are either minimal or overly demanding. Consequently, the lost child often withdraws from family interactions, which serves to perpetuate this dynamic. Such children tend to develop a high degree of independence and may internalise specific family expectations. The lost child may also assume other roles within the family, such as scapegoat or peacemaker. For instance, when occupying both the lost child and scapegoat roles, individuals may develop patterns of self-blame or diminished empowerment in relationships, alongside difficulties expressing emotions or forming a distinct identity. In adulthood, those who identify as former lost children frequently encounter challenges in establishing close relationships; they may avoid conflict and suppress emotions, complicating interpersonal connections. Within friendships, they might struggle to assert themselves, sometimes resulting in feelings of isolation. In professional settings, they often favour independent tasks but may experience difficulty with collaborative work due to concerns about their contributions being undervalued. The invisible child may transition between roles depending on environmental and relational contexts; for example, someone who was previously a scapegoat within their family may adopt the role of hero among friends or colleagues. Understanding these shifts provides valuable insight into the ongoing impact of formative experiences on current behaviour and relationship patterns.

PART THREE

UNDERSTANDING THE BOND BETWEEN ABUSER AND VICTIM

"Why would anybody go right back to the same place that they are going to get hooked, caught, and abused? Trauma bonding! It's loyalty to a person that is toxic/destructive"

Increasing awareness of mental health issues has resulted in more discussion about trauma bonds. Oversimplifying these concepts can result in misunderstandings that may hinder effective support for those seeking help. Trauma bonds can maintain cycles of abuse and explains why overcoming such patterns can be difficult.

Before defining trauma bonding, it is important to clarify what it does not describe. Trauma bonds do not refer to connections formed through shared hardship: While mutual understanding of difficult experiences can lead to supportive relationships or communities, a trauma bond specifically refers to an unhealthy attachment between an abuser and a victim, maintained by ongoing patterns of abuse and manipulation. Differentiating between these types of bonds can assist all involved in accessing appropriate resources and support.

Chapter Twelve

TRAUMA BONDING

"Trauma bonding is not love; it's an addiction to the narcissist"

Distinctions between trauma bonding and healthy relationships are evident in several domains including consistency, respect, trust, open communication, mutual empowerment, safety, and emotional well-being. Independence and clearly defined personal boundaries are central to differentiating these relational dynamics: A healthy relationship demonstrates consistent respect, trust, open dialogue, and encouragement of individual growth. Conversely, trauma bonding is characterised by fluctuating displays of affection, diminished trust, manipulation, a sense of powerlessness, and repeated infringements on personal boundaries.

Trauma bonding may develop during childhood resulting from recurrent cycles of abuse interspersed with minimal positive reinforcement, which cumulatively reinforce the bond over time. In the early stages of such relationships, individuals with narcissistic tendencies frequently exhibit affectionate behaviour. This initial kindness forms a normative basis for attachment between the perpetrator and the victim. Consequently, victims may find themselves emotionally attached to someone who initially displayed respect and warmth. If such behavioural patterns become normalised, victims are at increased risk of engaging repeatedly in relationships that mirror similar narcissistic or abusive cycles.

Abusive relationships often commence with intense affection and reassurances, commonly described as the honeymoon phase. The slow and gradual shift from positive to negative conduct may obscure recognition of abuse, making it challenging for victims to extricate themselves. Since victims often associate their feelings with the more favourable version of their partner, they may attempt to elicit the return of those preferred behaviours. Abusers typically perpetuate an emotional cycle to maintain control, ensuring that the victim remains emotionally invested and less likely to leave the relationship despite ongoing harm. Abusive partners may exhibit aggressive behaviour as a response to external pressures which might include health issues, fatigue, and work-related problems. Such situations can lead to frustration and dissatisfaction for victims of manipulative behaviours, which may contribute to feelings of powerlessness, injustice, anger, and mistrust. In efforts to reduce tension, the individual experiencing mistreatment might attempt strategies to ease interactions with their partner. These circumstances can result in ongoing anxiety, with the victim remaining alert to the potential needs or moods of their partner. To avoid conflict, they might be cautious about expressing their emotions and prioritising their partner's needs. This dynamic can gradually influence behaviour patterns within the relationship, resulting in difficult and complex considerations regarding living financial independence, and concerns for child safety and custody. Emotional ties may also exist, making the decision to leave more complicated.

Trauma bonding describes the attachment that can form between individuals in abusive relationships when they experience intense, emotionally charged situations together. This phenomenon holds similarities to the Stockholm Syndrome, where hostages develop connections to their captors. The combination of emotional closeness and risk may lead to strong bonds that appear irrational from an external perspective. In both cases, these emotional bonds can serve as coping mechanisms to manage stress and ensure self-preservation. It is, however, important to understand that the bonds formed during these extraordinary circumstances do differ. Trauma bonding is an ongoing abuse or mistreatment, which occurs inside a specific relationship involving a one-sided power imbalance. In contrast, The Stockholm Syndrome is often associated with short- or long-term hostage situations with externally imposed power balance due to physical captivity.

Trauma bonding can occur in cycles that include periods of intense conflict followed by a phase in which the abuser demonstrates affectionate or positive behaviour. Individuals who engage in such abuse may display traits such as kindness or righteousness and may seek to maintain contact with those they have harmed, even after separation. In some cases, the abuser may perceive themselves as having been wronged. The relationship can involve elements of exploitation, fear, and risk, which may result in significant emotional distress for the victim that further ties them to an abuser, even after separation. Intermittent reinforcement during psychological abuse involves alternating negative treatment with occasional positive gestures, such as giving gifts or taking someone out to dinner, which are provided unpredictably during the cycle. This pattern can lead victims to continually seek approval, often tolerating negative behaviour for intermittent positive interactions: A recurring dynamic, regardless of the context (for example, spousal or parent-child relationships). Occasional positive actions are often perceived as particularly meaningful due to their rarity, a phenomenon sometimes described as the 'small kindness perception'. In challenging or threatening circumstances, these rare positive experiences may provide hope that the situation will improve, which victims may focus on as a potential sign of change.

Abuse frequently hinges on tactics such as threats, manipulation, control, shaming, gaslighting, and sabotage, interspersed with periods of calm or displays of affection. These alternating patterns reinforce unhealthy attachments, perpetuating the relationship. It is also common for individuals to experience multiple abusive relationships; a repetition driven by the brain's tendency to seek out the perceived relief and safety associated with brief positive experiences amid ongoing abuse. Intermittent reinforcement is a powerful psychological mechanism—its impact can be observed in contexts such as gambling, where unpredictable rewards, like those from slot machines, motivate continued engagement.

Chapter Thirteen

LOVE BOMBING

"It's not always easy to see at first, because their gestures are meant to disarm you, to blow you away. But there is a difference between a caring person and a love bomber"

It can be challenging for individuals to differentiate between love bombing and a partner expressing genuine interest. Evaluating whether healthy boundaries are being respected may help identify possible concerns in the relationship. When one partner does not acknowledge the other's boundaries or if someone feels unheard or disregarded, it may indicate an issue. A person who is consistently argumentative or repeatedly disregards boundaries could be engaging in love bombing behaviours. Gaining insight from friends and family about the relationship can also provide helpful perspectives. Additionally, it is advisable for individuals to regularly reflect on their feelings and trust their instincts if something seems amiss. Love bombing refers to a pattern of psychological and emotional behaviour often characterised by excessive flattery and grand gestures. These actions can initially enhance the other person's sense of self-worth and make them feel valued. However, the underlying intention may involve seeking affection as well as exerting control. Over time, such gestures may be used to influence the other person's dependence in the relationship. Early signs can include intense experiences that lead to discomfort or feelings of being overwhelmed. Another indicator is missing time with family and friends, especially if spending time with others leads to conflict within the partnership.

Signs of love bombing include frequent and disproportionate expressions of admiration and commendation; persistent communication initiated by the narcissistic partner; giving unsolicited or unnecessary gifts; and premature and intense discussions regarding long-term plans for the relationship. It may occur between romantic partners, family members, or friends, and is often rooted in the abuser's insecurities and issues related to trust. This behaviour can be either intentional or unintentional. The individual's anxieties and insecurity may lead to an overdependence on the other person. Love bombing can result from learnt behaviours in childhood, such as exposure to passive-aggressive parenting, unresolved trauma, or experiences in previous relationships. Rejecting advances from someone engaging in love bombing might elicit threats or negative criticism, as the individual seeks constant reassurance of being loved and valued. Such individuals frequently attempt to accelerate relationship progression, swiftly labelling their partner as a soul mate or expressing fantasies about elopement or destiny. Introducing discussions around commitment early, bypassing significant milestones, and urging rapid movement toward a committed relationship typically reflect a desire to establish intimacy, closeness, and security prematurely.

Overcommunication of affection does not always manifest in conspicuous acts such as gift giving, grand gestures, or direct manipulation. It can often occur subtly within everyday interactions. A narcissist may frequently initiate contact when absent, repeatedly verbalise their feelings, or monitor the other person's whereabouts. Additionally, they may excessively post affirmation of their emotions online, in attempts to gain public validation of the relationship. Such behaviours can leave the recipient feeling overwhelmed, uneasy, or destabilised. It is reasonable for individuals in a relationship to assess whether both parties share similar expectations and levels of emotional investment. Each partner may experience and express love at varying paces and stages—what might feel appropriate to one person may be uncomfortable for the other. Open communication about these feelings should, ideally, restore a balanced dynamic. However, if such concerns are dismissed or invalidated, it may indicate the presence of love bombing and potential relational difficulties.

As the relationship progresses, a narcissist may become increasingly demanding of time, attention, and emotional energy. This can escalate into anger or jealousy directed towards the partner's friends or family members. Unfamiliar ultimatums might arise, pressuring the individual to choose between the narcissist and other significant relationships or responsibilities. The preferred outcome for the narcissist is monopolisation of their partner's time, fostering dependence. In doing so, the victim may become increasingly isolated and reliant on the narcissist. Addressing a love bomber's behaviour directly or establishing healthy boundaries may lead to defensiveness or argumentation. The individual may challenge the perspective of their victim and attempt to make them question decisions to decline their requests or propose changes. If a person experiences repeated boundary violations, it may indicate that their input is not being valued within the relationship. Conversely, individuals with narcissistic tendencies may exhibit anger and an unwillingness to accept rejection, often reacting negatively when their expectations are not fulfilled.

Individuals exhibiting narcissistic behaviours may attempt to isolate those around them from their social networks, thereby increasing their influence and control. Such isolation can be overt or, at times, subtle; for instance, they may display anger, distress, or sadness when others engage in activities independently. Coercing someone into actions that cause discomfort, or a sense of insecurity is indicative of emotional abuse. Love bombing tends to occur in three phases:

> *The idealisation phase.* The narcissistic partner frequently begins the relationship with pronounced expressions of affection and attention, fostering heightened trust and

emotional openness. Consequently, the other individual may find themselves rapidly drawn in or may view the relationship as unusually positive.

The Devaluation phase usually follows with the individual exhibiting narcissistic behaviour seeking frequent attention and time after a relationship becomes established. They could express discomfort if plans are made without their involvement. In some cases, they might restrict access to friends and family or attempt to convince the other person that certain actions are for their benefit. Methods such as fear and intimidation may be used to influence the other's behaviour, and in some instances, individuals could escalate to using violence.

The discard phase becomes apparent while addressing a partner about behaviour perceived as harmful, to establish clear boundaries, may lead to situations where the partner does not accept responsibility and is unwilling to cooperate. This can result in the other person feeling uncertain or unsettled about the outcome.

There are several significant signs that someone is love bombing. They include:

Giving needless gifts: An individual engaging in love bombing may present recipients with excessive or unsolicited gifts that are disproportionately lavish as demonstrations of their affection (for example, gifts that significantly exceed customary tokens such as flowers on a first date). If these gifts continue to be provided despite clear communication that they are unwelcome, this behaviour may serve as a strong indicator of love bombing.

They are in a rush to lock the relationship down: Rapid escalation is a characteristic often observed in individuals who engage in love bombing. They may quickly refer to their partners as soulmates, express fantasies about elopement, suggest that meeting the partner fulfils a lifelong aspiration, and seek commitment early in the relationship.

They are always available and demanding of attention: Individuals who engage in love bombing tend to exhibit a heightened dependence on their partner for emotional support, time, energy, and commitment. Over time, their behaviour may become increasingly demanding, often manifesting as anger or jealousy towards the victim's friends and family members. This can lead to unreasonable ultimatums that compel the individual to choose between the relationship and other valued relationships. In some cases, this behaviour may escalate to restricting the victim's ability to work, fulfil responsibilities, or pursue hobbies. The aim is to monopolise the victim's time and become their primary source of support. Such individuals may induce guilt to ensure they are prioritised over others.

They cannot take 'no' for an answer: The principle of consent is absolute: 'No' must always be respected. However, when one communicates to a love bomber that their behaviour is inappropriate, or attempts to establish healthy boundaries, it may lead to disputes wherein the individual's reasoning is challenged, potentially causing them to doubt their own perspective.

They prefer their victim alone: Limiting contact between individuals and their family or friends can increase one person's influence over another and affect their social activities. This may be done directly, such as requiring presence during all visits, or more subtly, for instance, by expressing negative emotions when certain expectations are not met. Attempts to direct or manage many aspects of a partner's life can indicate patterns associated with emotional abuse.

They over communicate their love: Love bombing does not solely involve excessive gift-giving, elaborate displays, or direct manipulation; it may also manifest in constant communication and a persistent desire to monitor the victim's location. Additionally, frequent public declarations of affection online can play a significant role.

They make their partner feel overwhelmed, uneasy, or off-balance: The development of love occurs at different rates and across various stages in new relationships. It is acceptable for both parties not to share the same sentiments simultaneously. If there are feelings of discomfort, imbalance, or that the relationship becomes overwhelming, it is advisable to carefully consider the dynamics of the relationship. Open communication regarding these concerns should be approached with understanding; however, if such expressions are not acknowledged or addressed constructively—as would be expected in a healthy relationship—this may indicate underlying issues.

Leaving an abusive relationship can involve complex and conflicting emotions. Recognising that a partner is not as previously perceived may be challenging, especially after forming attachments. These aspects of the relationship can make separation difficult. At this stage, individuals may be at risk of returning to the partner and continuing patterns of alternating abusive and positive behaviours within the relationship.

Recovery from experiences of love bombing within a couple depends on both parties. For the relationship to continue after concerns have been raised, it is important for all involved to recognise the need for new boundaries and acknowledge that certain behaviours are not acceptable. If both individuals are open to learning and making changes, there may be potential for progress. However, if discussions do not lead to positive change and problematic behaviours persist or escalate, considering how to safely conclude the relationship may be necessary. It is uncommon that significant behavioural change will occur in these circumstances. It can be helpful to seek support from family, friends, or qualified professionals to process and recover from the effects of love bombing. Regardless of the situation, it is important to separate responsibility for harmful behaviour from those who experience its impact. Victims of love bombing often experience depression, anxiety, anger, sadness, confusion, mourning, and loss. It is beneficial to mental wellbeing to obtain guidance from therapeutic sauces, who can offer validation of individual's experiences and emotions, assist in managing stress and emotional responses when appropriate, and provide guidance on ending unhealthy relationships. Individuals who feel unsafe or are experiencing physical violence or abuse are encouraged to consult their physician. While love bombing is a complex, and sometimes misunderstood phenomenon, it should be approached with seriousness. The aftermath of love bombing may result in challenges related to trust and forming new relationships. However, recovery is possible and begins by fostering openness, honest communication about past experiences, and establishing healthy boundaries with future partners from the outset.

Chapter Fourteen

TRUST AND DEPENDENCY

---

*"To know when to go away and when to come
closer is the key to any lasting relationship"*
Doménico Cieri Estrada

A position of trust refers to a relationship in which one individual possesses authority, responsibility, or influence over another person. This dynamic can lead to dependency and increased vulnerability for the individual subject to that authority. Such relationships are found in various settings, including caregiving, education, sports, and religious environments. Legal frameworks have been established to recognise these inherent power imbalances and to protect vulnerable groups from exploitation. Because positions of trust may create opportunities for abuse, individuals who hold these roles are bound by both legal and ethical obligations to use their authority responsibly. Recognising these positions is vital, as it highlights the significance of power dynamics and the necessity of safeguarding those at risk. This approach identifies the roles and relationships that require heightened attention to protective practices, contributing to safer environments for everyone involved.

The concept of the dependency paradox in interpersonal relationships explores the balance between healthy reliance on a partner or caregiver and the ability to use that relationship as a secure base to foster greater autonomy. Historically, in western societies, behaviourist perspectives from the 1940s encouraged parents to support self-soothing in children, suggesting that limiting emotional and physical support would promote independence and resilience. This philosophy has had a lasting impact, influencing adult romantic relationships by stressing the importance of self-fulfilment and cautioning against co-dependency. Excessive dependency in romantic partnerships is often viewed as harmful, with problematic outcomes such as substance abuse and domestic violence considered particularly damaging.

Attachment theory, however, argues that dependency is a natural aspect of human biology. Securely attached couples form psychological interdependence, a process that begins during pregnancy through the synchrony between mother and child. This biological connection supports optimal development, and after birth, reciprocal interactions—like shared smiles—help to build trust and a sense of security. Maternal actions, including physical contact and feeding, are crucial in regulating the infant's physiological states. These exchanges illustrate the reciprocal nature of attachment, which establishes the groundwork for healthy adult relationships. The release of neurochemicals such as oxytocin strengthens emotional bonds and reduces stress, thereby nurturing security and emotional wellbeing. Even when securely attached children experience distress during separation, they develop the confidence to explore their surroundings, knowing they can return to a stable source of support. Those who have secure attachments in childhood tend to demonstrate greater confidence and independence as adults, drawing reassurance from the presence of reliable attachment figures.

Behavioural patterns are formed early in life as adaptive responses to one's environment and often persist into adulthood. When these behaviours become maladaptive, understanding their developmental roots is key to effective intervention. Trauma can obscure the recognition of such patterns, sometimes sustaining cycles that shield individuals from confronting painful original experiences. Achieving meaningful change requires a thorough understanding of the motivations underlying recurring behaviours, which can be accomplished by analysing the self-protective cycles that arise in response to trauma.

Chapter Fifteen

CRITICISM

---

*"Your value does not decrease based on
someone's inability to see your worth"*

Frequent criticism from a partner, particularly one exhibiting narcissistic trait, may increase tension and cause individuals to feel that their efforts are inadequate. Ongoing criticism can contribute to feelings of discouragement and may lead to resentment. If criticism arises as an issue in relationships, it is important to address it before it escalates. Criticism can be direct, through comments or gestures meant to highlight faults, such as remarks about physical appearance or effort. Alternatively, passive-aggressive criticism may manifest through sarcasm or jokes that have underlying negative intent.

Criticism within relationships can create internal conflicts and negatively affect the partnership. Responding to negative remarks with similar criticism can perpetuate unproductive communication. A more effective strategy is to express how certain comments make affect feelings. Using 'I' statements reduces defensiveness and encourages constructive dialogue, whereas 'you' statements may increase negativity and defensiveness. By communicating personal feelings rather than focusing on perceived shortcomings, it is possible to approach conflict resolution more productively.

Criticism sometimes reflects underlying dissatisfaction in the relationship. Self-reflection on behaviours that could be improved may contribute to a healthier dynamic: If both partners consider changes in their behaviour there can be significant benefits to the relationship. Additionally, focusing on positive aspects and expressing appreciation is as important as discussing areas for improvement. Suggesting shared activities, and accepting openness to feedback as constructive criticism, can enhance the relationship further, aiding in fostering a healthier connection. In some situations, criticism may become abusive. If a partner regularly uses criticism to control or intimidate, this can be considered a form of emotional abuse.

Chapter Sixteen

GASLIGHTING (part 1)

"Gaslighters avoid responsibility for their toxic behaviour by lying and denying and making you question facts, your memory, and your feelings"
Karen Salmansohn

The term gaslighting originates from the 1938 stage play *Gas Light*, in which a husband seeks to undermine his wife's reality by subtly dimming their gas-powered lighting and then denying any change when she expresses concern. Gaslighting is recognised as a particularly insidious form of psychological abuse, leading victims to question their emotions, perceptions, and even their sanity. This manipulation enables the perpetrator to exert significant power and control, which are central elements in abusive dynamics. After eroding the victim's trust in their own judgement, the abuser increases the likelihood that the victim will remain within the abusive relationship, as leaving becomes more psychologically challenging. Two examples of phrases used in gaslighting are: *You're crazy, that never happened,* and *Are you sure? You tend to have a bad memory,"* and *"It's all in your head.* There are several techniques that Gaslighters might use:

Withholding: The abusive partner feigns misunderstanding or deliberately chooses not to engage in listening.

Countering: The abusive partner challenges the victim's recollection of events, even when the victim provides an accurate account.

Blocking or diverting: The abusive partner diverts the conversation or challenges the victim's perspectives.

Trivialising: One partner may dismiss or disregard the other person's needs or feelings.

Forgetting or denial: The abusive partner may deny the occurrence of events or assert that they do not recall them, including any promises that were made.

Individuals exhibiting gaslighting behaviours may utilise specific phrases to exert influence over others. Recognising these patterns and establishing new boundaries is essential for ensuring long-term wellbeing. The following statements are pointers to watch out for:

"This is not what happened."
"I never said that."
"You are being paranoid."
"It is not that big of a deal."
"You're imagining things."
"It's just a joke."
"Stop being so sensitive."
"You are clearly overreacting."
"You need to get your facts straight."
"I did not do that."
"You are making things up."
"Stop overthinking everything."

Gaslighting usually occurs in a gradual progression whereby the abusive partner's actions are harmless at first, causing them to go unnoticed. As patterns of abusive behaviour persist over time, individuals may experience confusion, increased isolation, and various adverse emotional effects. Additionally, the victim may increasingly question their perception of reality and become more dependent on their partner. Victims might: Frequently question memories and emotions; repeatedly ask *am I too sensitive*; throughout the day; often feel confused and experience a sense of instability; regularly apologise to the narcissist; have difficulty understanding the reasons for a lack of happiness despite positive life circumstances; make frequent excuses to friends and family regarding the narcissist's behaviour; withhold information from friends and family, so there is no need for explanation; have inability to express what is wrong, and do not accept something is wrong; lie to avoid the put-downs and reality twists; have trouble making simple decisions; have a sense that there have been to many inward changes, and that previous confidence, and the fun-loving side of them has vanished; suffer feelings of hopelessness and joylessness; feel that the ability to do anything right is lacking; and have doubt over being good enough.

Chapter Seventeen

ADDICTION TO THE CYCLE

"The worst part about anything that's self-destructive is that it's so intimate. You become so close with your addictions and illnesses that leaving them behind is like killing the part of yourself that taught you how to survive"
Lacey L.

A systematic analysis offers a comprehensive understanding of the discrete stages within the trauma bonding cycle. The following five phases typically characterise this cycle and serve to illustrate what individuals may experience or encounter as victims:

The tension building phase: During this phase, one party may become more irritable, controlling, or critical towards the other, leading to a tense atmosphere. The other party may try to prevent conflict by accommodating the other's wishes, which can result in heightened alertness and unease.

The incident or abuse phase: Tension increases, culminating in an incident where the perpetrator may engage in physical aggression, emotional mistreatment, or manipulative behaviour. The victim is subjected to mistreatment, resulting in feelings of distress, fear, and a sense of helplessness.

The reconciliation phase: After the abuse, the abuser shows remorse and promises change. Love bombing may occur, with the abuser showering the victim with affection, gifts and promises of improvement. The victim holds onto the hope that the loving partner they experienced during this phase will last.

The calm phase: The relationship enters a period of relative calm and stability. The abuser temporarily supresses their abusive behaviours, and the couple may enjoy moments of normalcy and harmony. The individual holds on to these calm periods, considering the possibility that the negative behaviour was an isolated incident.

An initiating event—such as a disagreement or an error attributed to the victim—can reactivate the cycle, leading to renewed tension and the subsequent repetition of the pattern.

In trauma bonding, individuals may develop a reliance on the cyclical patterns of abusive relationships due to intermittent reinforcement and the expectation that the relationship will improve. The reduction of tension and occasional expressions of affection can reinforce psychological attachment. People sometimes focus on actions or words as means to return the relationship to a previous positive state. Continued involvement in such relationships can be influenced by dependence on the cycle and apprehension about separation. For some, leaving may represent the loss of valued moments of affection. Additional factors such as concerns about loneliness or challenges in establishing new relationships can contribute to decisions to remain. Over time, these patterns can decrease self-esteem and self-worth, with repeated incidents reducing confidence and increasing feelings of powerlessness. Individuals may also internalise negative feedback from the other party.

Multiple factors can play a role in the development of addiction, which is characterised by a reduced ability to regulate or cease repetitive behaviours, leading to negative consequences. Traditionally associated with substances such as drugs, alcohol, gambling, and tobacco use, addiction may also pertain to activities including work, internet usage, inhalant consumption, shopping, or sexual behaviour. Furthermore, individuals may develop addiction to neurochemical responses elicited during cycles of abuse. Substances or activities that alter mental or physical states are often perceived as rewarding, which may encourage continued engagement. Gambling, for instance, can elicit comparable emotional responses and lead to repeated participation. Over time, higher levels or increased frequency of activity may be required to produce similar effects, contributing to difficulties in discontinuing the behaviour and intensifying cravings. Withdrawal symptoms may emerge when access is restricted.

Chapter Eighteen

LOSS OF SELF

---

*"Sometimes you need to lose yourself a few times
before you can find yourself"*

Self-loss refers to a profound and disorienting experience frequently described by survivors, marked by a significant reduction in their sense of identity, autonomy, and safety following abuse. While the term self-loss is commonly utilised, it is not entirely accurate; survivors are not fundamentally lost but have been exposed to circumstances that promote disorientation and detachment from their authentic identities. Recovery involves re-establishing agency and empowerment, which often includes building supportive relationships through reconnecting with family and friends or forming new support systems. The healing process requires survivors to develop insight into the nature and effects of the abuse, enabling them to regain control over their lives and bodies. Typically, this journey is protracted and complex, demanding patience, resilience, and a self-directed pace. The concept of *loss to abuse* can be analysed to better understand the effects of manipulative relationships:

Erosion of identity: Abuse can undermine an individual's sense of self, often resulting in feelings of shame, confusion, and a distorted self-perception.

Social isolation: Perpetrators may deliberately sever survivor's connections to their support systems, which can lead to marked loneliness and isolation.

Psychological impact: Survivors frequently experience intense emotions such as anger, sadness, fear, and anxiety, and may present with symptoms consistent with Post-Traumatic Stress Disorder (PTSD), including flashbacks and panic attacks.

Chapter Nineteen

RESIGNATION AND SUBMISSION

"Leaving a toxic relationship is not running away but self-preservation. It's the first step towards reclaiming happiness"
Donna Siggers

Resignation and submission are commonly identified elements of trauma bonding, in which individuals may experience helplessness and a tendency toward compliance within abusive relationships. Resignation denotes the individual's acceptance or surrenders to the abusive dynamics present. This state is often characterised by a perceived lack of power and a belief that they cannot alter their circumstances. Factors such as fear, learnt helplessness, and concerns that leaving may result in greater harm commonly contribute to resignation. Submission, on the other hand, involves the individual adhering to the abuser's demands, expectations, or mistreatment. Victims may suppress their own needs to prevent conflict or additional abuse and may prioritise the abuser's requirements over their own, causing profound effects to their wellbeing that are detailed below:

Cognitive dissonance: Individuals may have trouble reconciling inconsistent behaviour from the abuser, leading to confusion and cognitive dissonance.

Emotional effects: Trauma bonding can produce fluctuations in emotions, resulting in instability and emotional distress.

Reduced self-esteem and self-worth: Ongoing criticism from an abuser may be associated with lower self-confidence.

Increased dependency and concern about abandonment: Individuals may become reliant on the abuser for validation and develop concerns about being left if they attempt to exit the relationship.

Social isolation and alienation: Abusers sometimes separate victims from their support networks, which can result in reduced contact with friends and family.

Diminished autonomy and independence: Trauma bonding can lead to decreased control over personal decisions as the abuser exerts influence in different aspects of life.

Challenges in trusting and forming relationships: The dynamics of trauma bonding may contribute to difficulties in establishing trust and maintaining healthy interpersonal connections.

Higher risk of experiencing future abuse: Trauma bonding is linked to an increased likelihood of involvement in similar relationships.

PART FOUR

TYPES OF ABUSE

"Scar tissue is stronger than regular tissue.
Realise the strength"
Henry Rollins

Abuse refers to the use of behaviour or influence to inflict harm or exert control over another individual, particularly when such conduct is repeated. Many individuals may find it challenging to identify abuse, as it frequently originates from someone familiar and trusted. Periods of emotional difficulty can further impede the recognition of abusive patterns. Moreover, abuse often manifests gradually and imperceptibly, a phenomenon commonly associated with child and domestic abuse. Abuse may take various forms and occur in numerous contexts, affecting either oneself or others within one's personal network. Abuse may manifest in multiple forms, and individuals can be subjected to more than one type of abuse simultaneously. The following chapters outline differing aspects of abuse.

Chapter Twenty

PSYCHOLOGICAL ABUSE

"Just because it's explainable, doesn't mean it's excusable"

Psychological, or emotional, abuse refers to a pattern of behaviour where an individual uses insults, humiliation, or intimidation to exert control over another person. This type of abuse can affect the individual's perception of reality, sometimes leading them to internalise the negative messages. An isolated incident does not typically constitute emotional abuse; rather, it is the ongoing nature of such behaviours that creates an environment of fear and control. Emotional abuse can take place in various interpersonal settings, including parental, romantic, friendship, or professional relationships. Individuals who experience emotional abuse may encounter short-term effects such as confusion, fear, difficulty focusing, lowered confidence, nightmares, physical complaints, or increased heart rate. Over time, more persistent consequences can include anxiety, insomnia, and social withdrawal. Emotional abuse involves patterns of manipulation, control, or threats such as: Monitoring and regulating an individual's activities, including their social interactions or financial decisions; issuing threats to a person's safety, property, or loved ones; isolating an individual from family, friends, and support networks; engaging in demeaning, shaming, or humiliating conduct; exhibiting extreme jealousy, unfounded accusations, or paranoid behaviour; providing continuous criticism, employing frequent ridicule or mocking; making acceptance or care contingent upon specific choices; prohibiting an individual from spending time alone; obstructing an individual's professional or personal aspirations, inducing feelings of self-doubt and worthlessness; and causing an individual to question their competence or even their perception of reality.

Abuse results in profound negative experiences affecting day-to-day coping mechanisms. Victims might: Regularly experience belittlement or ridicule; develop a persistent sense of inadequacy; be unfairly held responsible for issues beyond their control; consistently feel the need to be cautious in their interactions; begin to doubt the appropriateness of expressing concerns or communicating with others; feel compelled to seek approval from the abuser before acting; and gradually withdraw from relationships with family and friends.

Emotional mistreatment and abuse can take multiple forms, occasionally being masked by considerate or affectionate language. When such abuse is subtle, it may not be immediately recognised. Alternatively, it may occur through extended periods of silence. Failure to identify these signs may prevent acknowledgment of the situation. Perpetrators might assert that the affected party is excessively sensitive, claim that such dynamics are typical of all relationships, or question the legitimacy of the victim's perspective. Such behaviours often include assigning blame rather than engaging in constructive resolution and may involve viewing the victim as subordinate. Furthermore, frequent sarcasm or attempts to regulate another person's emotional responses, though sometimes perceived by the perpetrator as helpful, constitute harmful conduct.

There are common emotional abuse tactics that are important to note: Constant criticism, shaming, humiliating, name calling, and manipulating; shouting; making accusations; jealousy, and possessiveness; sulking or refusing to communicate; monitoring whereabouts and communications; gaslighting that the victim's unhappiness is their own fault; isolating their victim from others; making the victim doubt their own sanity; threatening to destroy property; threatening to hurt their victim (physically or emotionally);

threatening to commit suicide; persuading the victim that it's not abuse; telling the victim that they deserve or have caused the abuse; and threatening to kill or harm their victim. Perpetrators of abuse often employ sophisticated manipulation techniques, which can make it challenging for victims to identify ongoing emotional and psychological abuse. These methods may result in victims attributing responsibility for relational issues to themselves. Such patterns tend to persist until the victim recognises the manipulative conduct, expresses an intention to leave, or terminates the relationship.

Chapter Twenty-One

GASLIGHTING (part 2)

"The Gaslighter avoids responsibility for their toxic behaviour by lying and denying and making you question facts, your memory, and your feelings"
Karen Salmansohn

Gaslighting is a form of emotional manipulation in which one party causes another to question their own perception of reality. This dynamic can result in the manipulator gaining increased influence, leading the affected individual to become reliant on them for validation. Gaslighting may contribute to symptoms such as confusion, self-doubt, anxiety, and depression. Individuals displaying narcissistic tendencies may exhibit possessiveness and heightened sensitivity, which can foster controlling dynamics within relationships, including shifting blame for adverse actions onto others. These behaviours may be associated with underlying psychological issues, such as antisocial personality disorder or narcissistic personality disorder. In certain instances, emotional abuse can escalate to physical violence if non-physical methods of control prove ineffective. The potential for physical harm can engender apprehension or fear in those affected.

Someone who is gaslighting might: Feign a lack of understanding regarding their own actions; characterise their partner's thoughts as irrational or imagined; challenge their partner's accurate recollection of events; deny having made promises acknowledged by the other party; and minimise their partner's emotions by suggesting they are overly sensitive when the reactions are reasonable. Over time, these behaviours may lead victims to experience confusion, anxiety, isolation, and depression. Gaslighting constitutes a form of emotional abuse that undermines an individual's psychological wellbeing. Additionally, such conduct can adversely affect the perpetrator; if one partner experiences stress, withdrawal, or dissatisfaction, overall relationship quality declines for both parties. These dynamics impede the formation of healthy, intimate relationships. Dysfunctional patterns like gaslighting are frequently rooted in early life experiences, often modelled by parents or significant adults. For instance, children raised in environments with parental substance abuse may be compelled to conceal such issues through manipulation. Nonetheless, perpetuating maladaptive coping mechanisms from childhood into adulthood cannot be justified. It is essential to acknowledge that individuals may remember events differently; variations in memory are commonplace and warrant respect. Addressing disagreements with humility minimises potential harm and cultivates greater empathy and understanding within relationships.

Chapter Twenty-Two

COERCIVE CONTROL

"Coercive control can take many forms but the goal of an abuser using coercive control is to intimidate their victim and limit their freedom"

Coercive control describes a recurring pattern of behaviours, including, but not limited to assault, threats, humiliation, and intimidation, intended to influence or regulate another individual's actions. This conduct is classified as a criminal offence when it arises within intimate or familial relationships, whether ongoing or concluded. Examples may include managing another person's finances or restricting their interactions with friends and family.

Perpetrators of coercive or controlling behaviour often exploit the vulnerabilities and insecurities of others. Individuals subjected to such conduct may experience fear, confusion, and heightened intimidation. The consequences of these behaviours can affect multiple aspects of an individual's life, leading to diminished self-confidence and altered self-perception. Notably, coercive control and associated behaviours may continue even after the relationship has ended.

In coercive relationships, an abuser might: Use consistent criticism and humiliation techniques to undermine a victim's sense of self and ability—including their role as a partner or parent; display extreme dominance, demanding obedience and exhibit a sense of 'entitlement'; have pathological jealousy; isolate their victim from supportive systems such as friends, family, religious communities, and other peer groups alongside the interception of messages and telephone calls; prevent their victim from fulfilling their basic needs such as sustenance and shelter; monitor and control time and activities; track their victim online; control sleeping patterns and select their wardrobe; consistently reinforce feelings of worthlessness in their victim; force the involvement of their victim in criminality, including use of illicit substances and shoplifting. Subsequently using these acts as leverage through blackmailing or threatening behaviour; ensure their victim is financially stunted to control spending and their sense of independence; restrict the freedom of movement, and deny them autonomy; use threats of violence; threaten suicide, femicide, homicide, or familicide ensuring their victim believes they will act; use children to exert control through criticising or belittling the victim in front of them, or by threatening to remove the children from the family home; threaten to cause damage to property; threaten to maim pets; make threats to publish or share their victim's private information—this can include images; make false allegations against their victim; prevent their victim from accessing transport or work; control or monitor their victim's health and body, including preventing access to care and taking necessary medication; manipulate information given to professionals and agencies such as the family GP's or the police; coerce their victim into sexual relations, including making them feel that they owe sex to them, and anyone else of their choosing; and assault or rape their victim.

Coercive control frequently emerges gradually within interpersonal relationships, with perpetrators not necessarily demonstrating controlling behaviours consistently. Consequently, identifying signs of coercive control may prove challenging for both those directly impacted and external observers. Individuals subjected to such control might exhibit withdrawal or apprehension and may not always recognise their own victimisation. It is common for abusers to initiate, maintain, or escalate coercive and abusive actions even after the relationship has ended. When engaging with individuals suspected of experiencing coercive control, it is essential to approach the situation with sensitivity and professionalism. Victims may lack understanding of their circumstances resulting from being manipulated by the perpetrator. Creating a supportive and respectful environment can facilitate open communication and self-reflection, which may empower individuals to recognise their situation and pursue appropriate assistance. In supporting a victim of coercive control, the following steps should be considered:

Preparation: Identify safe methods for contacting the victim and be aware of available local resources. Keep contact details for national helplines accessible.

Initiate communication: Begin conversations in a considerate manner by expressing concern, such as asking: 'You haven't been in touch, is everything okay?'

Listening: Providing a receptive space for the victim to speak can facilitate disclosure, even with limited knowledge of coercive control or domestic abuse. Opening discussions could include saying 'I can see there's something on your mind, can I help?'

Validation: Affirming the victim's experiences when they share accounts of coercive control or domestic abuse can contribute to more positive outcomes. Informing them that these behaviours are not acceptable may also be supportive.

Provide assistance: Offer suggestions rather than directives, as applying pressure may cause the victim to withdraw. Allow them time to make their own decisions. Useful assistance might be: Offer to call a helpline, suggest making a safety plan together, and to offer temporary accommodation or to keep an emergency bag (departing from an abusive relationship may present safety risks).

Seeking assistance from domestic abuse charities or the local authority is one way to begin reporting concerns regarding coercive control. When children are involved, it may also be necessary to refer the matter to children's services. Vitally, there should be understanding that individuals affected by coercive control might have difficulty articulating their experiences or understanding the potential risks involved.

Keeping a record of events to document evidence can be helpful if it does not compromise safety. In some cases, individuals suspected of abuse may attempt to influence professionals, which can further isolate those experiencing coercive control.

Chapter Twenty-Three

PHYSICAL ABUSE

"We do learn so much about ourselves in our experiences. But also, know that it shouldn't have happened. This was not a lesson you needed to learn"
Jordan Pickell

Physical abuse is defined as the deliberate infliction of injury or trauma through physical contact. It encompasses the intentional use of force causing harm, injury, or bodily trauma to an individual. Such acts may occur as isolated incidents or may be repeated over time. Being on the receiving end of negative and deviant behaviours can cause philosophical affects in victims. They might: Display apprehension about physical contact; present with unexplained bruises or marks they are reluctant to discuss; appear anxious in the presence of certain individuals; exhibit flinching or startle responses to sudden movements; and consistently demonstrate hypervigilant behaviour. Physical abuse is frequently utilised as a means for one individual to exert power and control over another. The early identification of signs indicative of physical abuse is critical for prompt intervention and the delivery of appropriate support services. Examples include:

Physical punishment: Physical punishment, also known as corporal punishment, involves disciplinary practices that use physical force with the intent to cause pain or discomfort. This encompasses actions such as smacking, slapping, striking, or shaking, as well as any use of force for punitive purposes. Such actions can result in various injuries, including bruises, lacerations, or fractures. Individuals who experience physical punishment may develop increased anxiety, social withdrawal, reduced self-esteem, and heightened aggression.

Strangulation: Strangulation is defined as the external application of pressure to the neck, leading to restricted blood flow or airway access and a consequent decrease in oxygen to the brain. Non-fatal strangulation refers to incidents where the act does not result in death. The intentional infliction of strangulation or impairment of breathing constitutes an offence, regardless of consent if serious harm occurs. Such conduct may involve reckless behaviour and carry legal consequences, including imprisonment.

Scalding and burning: Scalds are caused by exposure to hot liquids or steam, whereas burns result from contact with dry heat sources such as irons or open flames. These injuries can compromise multiple skin layers, potentially resulting in shock, heat exhaustion, or infection.

Restraint: Restraint describes measures or conditions enacted to exercise control over an individual, often for disciplinary or safety-related reasons. The improper or unlawful use of restraint may involve methods such as securing an individual to a chair or gripping their wrist to restrict movement, and may breach ethical and legal standards.

Deprivation of comfort: The deliberate infliction of discomfort, including removing blankets or opening windows during cold weather, can lead to adverse outcomes such as physical distress, sleep disturbance, impaired thermoregulation, and emotional harm.

Involuntary isolation or confinement: Restricting an individual's freedom of movement, particularly through confinement in enclosed spaces, may result in significant psychological consequences, notably increased anxiety, and cognitive decline.

Misuse of medicine: Misuse of medication encompasses excessive sedation, administering overdoses, or providing medication prescribed to others, each of which poses health risks.

Forcible feeding or withholding food: Forcible feeding includes interventions such as enteral (tube) feeding where medical intervention is required. Food is delivered directly into the stomach via a tube inserted through the nose or mouth. Improper use of enteral feeding is abuse. Cupping is an alternative method of forcible feeding, whereby liquefied food is administered using a spout, introducing risks such as choking or aspiration. Withholding food can quickly lead to nutritional deficiencies, weakened immunity, increased susceptibility to illness, severe weight loss, tissue wasting, and in extreme cases, life-threatening starvation.

Chapter Twenty-Four

DOMESTIC ABUSE

> *"There is no life to be found in violence. Every act of violence brings us closer to death. Whether it's the mundane violence we do to our bodies by overeating toxic food or drink, or the extreme violence of child abuse, domestic warfare, life-threatening poverty, addiction, or state terrorism"*
> Bell Hooks

Domestic violence is defined as the use of emotional, psychological, sexual, or physical force by one family member or intimate partner to assert control over another. The perpetrator may be a romantic partner, parent, or other relative, and these behaviours typically take place within the home environment. Violent acts include verbal, emotional and physical intimidation; destruction of the victim's possessions; maiming or killing pets; threats; forced sex; slapping; punching; kicking, choking, burning; stabbing; shooting; and murder. Those at risk of being targeted might be spouses, parents, stepparents, children, siblings, elderly relatives, and intimate partners. A victim may develop feelings of affection toward their abuser, which may result in confusion and the perception that such behaviour is acceptable; however, it should be noted that abuse is not considered justified under established guidelines. Individuals might: Experience persistent tension or anxiety within the home environment; make efforts to avoid provoking certain reactions from others; overhear raised voices or sounds of objects being broken; witness family members being harmed or sustain harm themselves; and be restricted from communicating with others, outside the household. Those with elevated narcissistic traits may demonstrate increased sensitivity and a higher likelihood of aggressive responses to provocation. Studies suggest that narcissism can act as a risk factor for aggression and violence. These individuals might also display unprovoked aggressive behaviour, which is not necessarily directed at specific targets. If concerning behaviours are observed in someone with pronounced narcissistic traits, it is recommended to monitor the following actions carefully:

Entitlement to special treatment: When individuals with narcissistic traits do not receive the level of favour they anticipate, they may respond with anger and, at times, display reactive outbursts. Their aggression can also manifest in calculated, detached, and proactive manners.

Feeling humiliated, criticised, or disrespected: This can undermine a narcissistic person's sensitive self-concept and cause feelings of insecurity. They may react defensively, employing a highly uncompromising stance as if protecting their core identity.

Inflated sense of self: In such cases, those with narcissistic traits may exhibit anger directed toward those they believe have caused inconvenience or frustration. This can include bystanders not involved in the situation.

Viewing others as inferior: Narcissists may perceive others as inferior, which can contribute to their sense of justification for emotional or physical mistreatment. They might believe that external circumstances are adversarial, leading them to view their actions toward others as a form of self-defence or necessary response.

Self-image: Individuals exhibiting narcissistic traits often seek to reinforce their self-esteem by acquiring symbols associated with superiority, including wealth, power, status, attention, or control. To enhance their image, they may employ various strategies to achieve these ends.

Chapter Twenty-Five

FINANCIAL ABUSE

"Part of the healing process is sharing with other people who care"
Jerry Cantrell

Financial abuse is recognised in law as a type of domestic abuse. It takes place in intimate partner relationships, caregiver relationships and within families where one person limits another's access to education, financial resources, spending, transportation, or technology, which can impact employment and social interaction. Financial abuse may also involve damage to property or not contributing to household expenses. This conduct falls under coercive and controlling behaviour and may result in long-term consequences. Financial abuse often occurs in conjunction with other forms of abuse, such as physical, sexual, or psychological. Its purpose is, typically, to cause financial dependence and instability, which can make it challenging for the affected individual to leave the relationship. Signs of financial abuse include unexplained money shortages, unusual financial transactions such as mysterious withdrawals, or a sudden inability to pay bills; lack of access to accounts with another individual controlling all monies, bank accounts and ensuring external control over spending and financial decisions, including forced spending on others; changes in living standards; missing essential items; and changes to important documents such as wills or deeds. Financial abuse also involves being pressured to make loans or sign over property; overdue bills, such as rent or utilities that have been unknowingly placed in your name, or finding someone else has added themselves as a signatory to personal bank accounts; and ensuring dependency by placing all assets in the abuser's name. Financial sabotage can be hard to recognise, yet it poses major challenges for those trying to regain independence after leaving an abusive relationship.

Chapter Twenty-Six

SEXUAL ABUSE

"We must send a message across the world that there is no disgrace in being a survivor of sexual violence—the shame is on the aggressor"
Angelina Jolie

Sexual abuse encompasses any form of sexual contact or behaviour that takes place without mutual consent. Such incidents may occur online, via text messaging, or over the telephone. It is important to note that prior consent, an absence of verbal refusal, or a subsequent change of mind does not constitute valid ongoing consent when the conduct is unwelcome. Unsolicited sexual contact can result in discomfort or fear. Individuals subjected to such acts may be coerced into sharing explicit images or viewing sexual content, potentially leading to emotional distress, confusion, or self-blame. Victims are often instructed not to disclose the incident(s). Narcissistic sexual abuse may involve attempts to control behaviour, reinforce a sense of superiority, pursue personal fantasies, or undermine the autonomy of another individual. While not all individuals with narcissistic traits engage in sexual dominance, understanding potential indicators can help clarify how such behaviour might develop:

The early stage: Abuse might begin with the narcissist grooming their victim with a mildly abusive act. They might demand sexting while they are at work, for example. Unwanted or embarrassing sexual acts are designed to catch a person off-guard and create a feeling of trepidation. Additionally, it is their way to send a message to others that they belong to them. This isn't meant to be comforting but, instead, something that leaves their victim feeling as if they are a possession. Sometimes narcissists share narcissistically obtained photographs with friends to add further humiliation. When confronted, a narcissist will minimise, deny, or blame the victim.

Verbal assaults: Verbal remarks may begin as compliments, potentially leading the recipient to feel appreciated by the narcissist. If the recipient does not comply to expectations, there may be expressions of disagreement related to sexual orientation. The narcissist might then accuse the recipient of manipulation or control and may openly critique their interests. Comments can sometimes become vulgar and include personal insults or negative statements about appearance, which could contribute to feelings of inadequacy. Narcissists engaging in sexually abusive behaviours as a form of control, often regard victims as means to fulfil their own needs rather than as individuals with emotions.

Jealousy rages: Jealousy rages may manifest as a request for details about previous relationships, which can then be referenced in discussions about current interactions or decisions. If jealousy arises in response, responsibility for this emotion may be attributed to the other person. In some situations, individuals might express preferences regarding public attire, sometimes asking others to dress more conservatively or provocatively. These choices can lead to further accusations, including suggestions of attraction to others, or infidelity which may be used to justify subsequent actions.

Coercion tactics: Some individuals may use manipulation strategies—such as harassment, blame, or guilt—to pressure a partner into sexual activity, without acknowledging this behaviour as abusive. They might request sex following conflicts or employ emotional approaches to obtain agreement, occasionally preventing their partner from leaving until consent is given. After compliance, the individual may withdraw or suggest motivations beyond intimacy.

Threatening infidelity: In certain cases, infidelity can be utilised as a manipulative tool if an intimate partner does not comply with increasing sexual requests. A person may demonstrate flirtatious conduct with others in front of their partner to influence or coerce them into participating in undesired sexual activities. Furthermore, indicating attraction toward a partner's acquaintances can function as a means of fostering social distance or isolation for that partner. If verbal strategies to achieve isolation prove ineffective, infidelity may subsequently be employed to reinforce this pattern.

The pushy phase: Repeated demands regarding sexual frequency or style are continually escalated. When the individual attempts to establish boundaries, the narcissist consistently pressures them to go beyond these limits. Any expression of dissent is met with ridicule or verbal confrontation, often leading the victim to acquiesce. Subsequently, instances of objection may be cited as justification for further punitive actions toward the individual.

Inciting fear: Submitting to unwanted sexual acts can result from incited fear by the narcissist to meet their sexual wishes. Threats may involve violence, abandonment, humiliation, punishment, betrayal, or financial abuse. The partner may reinforce these fears through actions, followed by pressuring the victim to engage in sexual activity as a demonstration of loyalty.

Selfish appeals: Engaging in unprotected intercourse can exemplify self-centred decision-making. In abusive relationships, sexual activity may focus solely on the preferences of an individual exhibiting narcissism. Such individuals might decline to use condoms and shift the responsibility for birth control or STD/STI prevention onto their partner. It is also possible for them to misrepresent their health status regarding STDs/STIs, avoid medical screening, and attribute any resulting issues to their partner. Concerns about unprotected sex are often dismissed or downplayed, as the emphasis remains on satisfying their own interests.

Sexual withdrawal: In certain relationships, a partner may choose to abstain from sexual activity. Requests for intimacy can be met with feedback regarding performance or various reasons for abstinence, and responsibility for reduced desire may be attributed to the other partner. There may also be periods alternating between increased intimacy and withdrawal that affect relationship dynamics. One partner might make decisions about the other's body or set expectations related to physical appearance, such as changes in weight, exercise, or grooming habits. Additionally, one partner's preferences may impact decisions during challenging circumstances, including matters related to pregnancy, abortion, or breastfeeding, rather than reaching a mutual agreement.

Violating ethical boundaries: In certain circumstances, an individual may seek to impact another person's values or personal limits by introducing a third party into intimate situations. This might entail inviting their partner to observe, or be observed, during acts that constitute infidelity.

The violent stage: In certain cases, when an individual's behaviour escalates to violence, sexual activity may cease to represent mutual affection or commitment. At this stage, acts of intimacy may be motivated by control or dominance rather than emotional connection. It is important to recognise that not all individuals with narcissistic traits reach this level; many remain at less severe stages of behaviour. Individuals who do escalate may participate in or request actions that are considered criminal under the law, irrespective of their relationship status. Marriage does not preclude the legal recognition of a person as a victim of a sex crime.

Sexual offences include non-consensual crimes such as:

Rape: Intimate relations require the consent of all parties involved. Legally, non-consensual penetration with a penis is classified as rape and applies to the vagina, mouth, or anus. According to legal definitions, only individuals with a penis can be charged with committing rape; however, others can be held criminally responsible if they assist or facilitate such acts.

Sexual assault: Where one person intentionally touches another person intimately without their consent. Touching can be done with any part of the body or with an object.

Assault by penetration: Intentional penetration of another individual, whether with the penis, any part of the body, or an object, without the individual's consent constitutes an unlawful act.

Sexually orientated crimes against children (children are defined as being under the age of eighteen) include:

Indecent images: It is prohibited to take, allow to be taken, create, possess, display, distribute, or publish an image of a child depicted indecently, including in a sexual context. This also applies to images where adults engage in sexual acts, and a child is present but not portrayed sexually. Images include actual photographs or video footage; images created digitally; making or downloading a photograph or video; printing an indecent image; and opening an email attachment containing an indecent image

Child sexual abuse: When children are compelled or encouraged to engage in sexual activity, regardless of their awareness, it is classified as child sexual abuse. Incidents may sometimes involve significant violence. Child sexual abuse might involve: Physical contact including rape or oral sex; non-penetrative activities include masturbation, kissing, external rubbing, and touching over clothing; non-contact abuse can be coercion into viewing or producing sexual images, being forced to watch sexual activities, encouraged to behave in sexually inappropriate ways, or exploitation/grooming in preparation for abuse (including via the internet), or prostitution. Adults and children of all genders can commit child sexual abuse

Prostitution: Although many aspects of prostitution are illegal, the act of exchanging sex for money or goods is not. Offences include persistently seeking sexual services; paying for sexual services in public; operating or owning a brothel; advertising prostitution; and engaging in kerb crawling for prostitution.

Extreme pornography: Extreme pornography describes pornographic images that are grossly offensive, disgusting, or obscene, and portray a range of extreme acts in an explicit and realistic way. This can include images of extreme violence; mutilation; sexual activity with an animal intended to sexually arouse; and disclosing private sexual images or films without consent of the people appearing in them.

Degrading acts: Persons exhibiting narcissistic characteristics may not recognise their behaviours toward others as demeaning. These actions are often intended to undermine and foster a sense of entrapment for the individual within the relationship. For instance, victims may encounter disparaging comments that question their value.

Sadistic sex: There are two forms of sadistic intimacy: Mild (also known as S&M) and severe which can lead to death.

> *Mild:* Individuals exhibiting narcissistic traits may partake in behaviours such as engaging in master-servant dynamics, the use of substances to limit mobility, infliction of pain (e.g. whipping), confinement within enclosures, physical restraint, blindfolding, or applying clamps to sensitive body areas. In the absence of explicit consent from all parties, these actions are deemed abusive.

> *Severe:* Individuals displaying narcissistic tendencies may exhibit persistent demands for severe behaviours, such as physical assault, psychological abuse, acts involving burning, cutting, stabbing, vampirism, or even homicide. Such behaviours may occur prior to, during, or following the primary incident.

The exit stage: Individuals may elect to terminate the relationship at any stage referenced above. Certain experiences can result in reluctance to disclose reasons for departure, often due to feelings of embarrassment or humiliation. Engaging with professional support services may facilitate the recovery process. The impact of sexual abuse may become more evident when an individual commences a healthy sexual relationship.

Post relationship: Following the end of a relationship with a narcissistic partner, two behavioural patterns are frequently noted. The individual may persist in perceiving an ongoing association with their former partner, which can manifest in continued requests for intimacy irrespective of new relationships. Conversely, they may behave as though the prior relationship never existed, dismissing any previous bonds. Individuals may alternate between these behavioural responses

Chapter Twenty-Seven

ONLINE ABUSE OR CYBERBULLYING

"Revenge is surviving, getting out, and being a better person than you were, and breaking the cycle"
Kirsty Green

Online abuse refers to injurious conduct executed via internet-enabled technologies. This behaviour can occur on any web-connected device, including mobile phones, tablets, and computers. Incidents may take place across multiple digital platforms such as social media, text messaging, messaging applications, online chats, email, online gaming, and live-streaming sites. Perpetrators may be known to the victim or complete strangers, and the abuse may transpire exclusively within digital contexts or extend offline through actions like bullying or grooming initiated through online interactions. Types of online abuse include:

Cyberbullying: Cyberbullying, sometimes referred to as online bullying, is a form of repeated harassment that occurs over digital platforms. Individuals experiencing cyberbullying may feel as though they are under ongoing observation and lack privacy.

Emotional abuse: Emotional abuse is defined as the persistent emotional mistreatment of an individual, occurring in both digital and physical contexts.

Exploitation: Sexual exploitation is defined as a form of sexual abuse where an individual is encouraged or coerced to produce sexually explicit images, videos, or engage in sexual conversations online.

Grooming: Grooming involves the deliberate development of a relationship with children or vulnerable individuals, with the intention of perpetrating abuse, exploitation, or trafficking. This behaviour can occur both online and offline and may be carried out by strangers or acquaintances of the victim.

Sexting: Sexting refers to the transmission of explicit adult content, such as sexual text messages, images (including semi-nude or fully nude photographs), videos, or unambiguous text, which may involve oneself or others. In certain circumstances, individuals may be pressured or coerced into sending these types of communications.

Sexual abuse: Sexual abuse is defined as circumstances in which a child, young person, or adult engages in sexual activities without providing informed consent. Such abuse can also take place online, including being compelled to view or distribute images or videos, or to participate in sexual acts during virtual interactions.

It can often be challenging to identify when an individual is experiencing online abuse; however, certain behavioural changes may provide important indicators, such as: Someone might spend a lot more or a lot less time than usual online, texting, gaming, or using social media; seem distant, upset, or angry after using the internet or texting; be secretive about who they're talking to and what they're doing online or on their mobile phones; and have lots of new phone numbers, texts, or email addresses on the devices. The consequences of online abuse or cyberbullying include: Receiving threats, nasty or overwhelming messages that include upsetting content; feeling left out of group chats or conversations on purpose; have their photos, videos or personal information shared without consent; being mocked, lied about, or targeted online; and feeling afraid to go online or that they are being watched or followed

Chapter Twenty-Eight

DISCRIMINATORY ABUSE

"Here bring your wounded hearts, here tell your anguish; Earth has no sorrow that Heaven cannot heal"
Thomas Moore

Experiencing unfair treatment or being singled out due to personal characteristics constitutes discriminatory abuse. Such actions may be based on age, race, gender, sexual orientation, disability, or religion, and can be perpetrated by anyone, including acquaintances, relatives, or members of the wider community. Discrimination might mean that a person gets called names or slurs, are excluded or treated differently, feel unsafe for being themselves around certain people, or be told they are oversensitive or imagining the abuse. In the United Kingdom, discrimination against individuals based on specific characteristics is prohibited by law. Those characteristics are age; gender reassignment; marital status, including civil partnerships; Being pregnant or on maternity leave; disability; race, including skin colour, nationality, or a person's ethnic or national origin; religious belief; and, sex and sexual orientation. Additionally, there are settings in which people are legally protected from discrimination. They are: In the workplace, or within education; as a consumer; while using public services; when buying or renting property; and as a member, or guest, of a private club or organisation. Examples of discrimination are:

Direct discrimination: Direct discrimination occurs when an individual with a protected characteristic is treated less favourably than others.

Indirect discrimination: Indirect discrimination occurs when policies or practices are implemented uniformly but disadvantage individuals with protected characteristics by placing them at an unfair disadvantage.

Harassment: Harassment constitutes unwelcome conduct related to a protected characteristic that undermines an individual's dignity or results in a hostile environment for them.

Victimisation: Victimisation is defined as subjecting an individual to different treatment because they have lodged a complaint about discrimination or harassment, or when they are consistently targeted for hostile behaviour.

Additionally, individuals with disabilities are entitled to the same rights as all other persons. Employers in the United Kingdom have a legal obligation to implement reasonable adjustments to support employees or applicants with disabilities. These adjustments may include application forms in Braille or audio; extra time for aptitude tests; accessible interviews (e.g. wheelchair access, and communicator support); appropriate workplace facilities and equipment; pay and employment terms; and access to work-related benefits, such as recreation or refreshment areas. Addressing discriminatory abuse typically starts with informal efforts, such as communicating directly with an employer. An impartial mediator can help facilitate a resolution. If these steps do not lead to an agreement, it may be useful to consult a third party like the HR department or Citizens Advice for further guidance.

Chapter Twenty-Nine

NEGLECT

"If you carry joy in your heart, you can heal any moment"
Carlos Santana

Neglect is defined as the ongoing failure to fulfil an individual's essential physical and psychological requirements. This can result in significant harm or hinder both health and developmental progress. Neglect may take place in care homes, family residences, or any environment where individuals depend on others for support. Individuals responsible for neglect may lack the capacity for self-care or fail in their duty of care toward others. Contributing factors may include carelessness, indifference, reluctance, or other forms of abuse. A person might: Be left alone in situations where safety cannot be ensured; lack access to adequate food, clean clothing, or necessary care; experience limited attention or interaction from others; face health issues that are not addressed appropriately; or have their requests for assistance or needs disregarded. Dividing neglect into separate components allows individuals and professionals to identify risk factors, understand underlying causes, and apply specific interventions known as *Omission, not commission:*

Neglect: Neglect is characterised by the omission of necessary caregiver actions, rather than by specific active behaviours.

Basic needs: The omission of essential provisions, including adequate food, shelter, clothing, and necessary medical care.

Psychological needs: The inability to fulfil emotional needs, including the provision of nurturing, stimulation, and security.

Long-term effects: Neglect can result in significant and enduring adverse effects on behaviour, academic performance, emotional wellbeing, and physical development. It impacts children, adults, and includes individuals residing in care facilities.

Risk to life: In extreme circumstances, neglect may result in significant harm or fatality.

Effectively addressing neglect is essential for preventing additional trauma to affected individuals. This can be done through:

Early intervention: Early identification and intervention for neglect can help reduce its potential effects.

Support for families: Support and resources are available to assist families experiencing challenges with parenting, as well as those caring for vulnerable individuals, to help prevent neglect.

Child protection: Identifying and reporting neglect is important for ensuring the wellbeing of vulnerable individuals.

Chapter Thirty

ORGANISATIONAL OR INSTITUTIONAL ABUSE

"A narcissist paints a picture of themselves as being the victim or innocent in all aspects. They will be offended by the truth. But what is done in the dark will come to light. Time has a way of showing people's true colours"
Karla Grimes

Organisational or institutional abuse occurs when there is maltreatment of a person from a system of power. Acts can range from child abuse, neglect, emotional abuse, physical and sexual abuse, and hunger. The abuse can sometimes be a harsh or unfair way of behaviour modification. An employee of the facility usually causes maltreatment. Therefore, abuse might occur in nursing homes, schools, hospital settings (both acute and in-patient), and prisons causing vulnerability to children, adults with learning difficulties, adults with mental health problems, older people, and prisoners. There are several categories of institutional abuse:

Overt abuse: Overt abuse may involve physical, sexual, or emotional harm inflicted by an individual employed at a facility within a structured system of authority.

Programme abuse: Programme abuse refers to circumstances within an institutional environment where a programme operates below acceptable standards or improperly exercises authority to influence individual's behaviour.

System abuse: System abuse refers to a care system operating beyond its capacity, which results in insufficient resources and subsequent inadequate treatment.

A person on the receiving end of abuse might be treated as if they are the problem; not have any say in decisions about their care; feel ignored or spoken down to; feel as if they're just another number; feel unsafe within the organisation or institution; or have their right to confidentiality disrespected.

Chapter Thirty-One

FEMALE GENITAL MUTILATION (FGM)

"Domestic violence is an epidemic, and yet we don't address it. Until it happens to celebrities"
Nelsan Ellis

Certain cultures, often for religious reasons, practice non-medically necessary genital alteration that involves deliberate modification or removal of tissue. This procedure is performed on children and women to control sexuality. There are numerous names for female genital mutilation, which include female circumcision or cutting. The practice can also be known as sunna, gudniin, halalays, tahur, megrez or khitan.

FGM is a form of abuse and is illegal in the United Kingdom. This practice is extremely harmful to both physical and emotional health. Indications that FGM might happen are: The victim learns about a ceremony during which they will 'become a woman' or be 'prepared for marriage'; a relative or someone known as a 'cutter' is planning to arrive from abroad; a female family member, perhaps their mother, sister or aunt, has had FGM; the family is planning a long trip overseas without explanation, especially during school holidays; or the victim is taken out of school unexpectedly or for a long period. FGM is a painful process that can result in the person suffering issues with walking, sitting, or going to the toilet; cause them pain, infections and difficulty during sex or childbirth;

experience flashbacks, anxiety, or depression; and feel the need to avoid doctors or medical appointments.

Chapter Thirty-Two

HONOUR BASED VIOLENCE AND FORCED MARRIAGE

"I'm still coping with my trauma, but coping by trying to find different ways to heal it rather than hide it"
Clemantine Wamariya

Honour-based violence refers to situations where an individual is harmed, threatened, or abused due to the belief within their family or community that they have brought shame or dishonour upon them. This may occur when the person does not adhere to specific expectations regarding behaviour, relationships, or identity. In some cultures, forced marriage forms part of their tradition. Forced marriage describes a situation in which a person is compelled to marry someone against their will, potentially involving threats, pressure, emotional coercion, or physical force. Both honour-based violence and forced marriage are prohibited by law in the United Kingdom. Victims can experience harm or disciplinary actions for not adhering to prescribed standards of behaviour, relationships, or identity; be subjected to monitoring or restrictions on movement and social interactions; lack autonomy in decisions regarding marriage or other significant life events; receive accusations of bringing disrepute to the family or community; and fear potential repercussions for expressing concerns or objections.

Chapter Thirty-Three

MUTUAL ABUSE

"The survivor would have nothing to fight back against if the abuser stopped abusing them"
Emma Katz

In abusive relationships, it is common for one individual to exhibit abusive behaviours while the other person responds accordingly. The notion of mutual abuse is a subject of debate, with some experts asserting that it may diminish the validity of experiences reported by those who endure abuse. Additionally, the term mutual abuse has been interpreted by some as an attempt to rationalise or legitimise certain actions within these relationships.

PART FIVE

THE EFFECT OF ABUSE ON MENTAL HEALTH

"Your mind is a powerful thing. When you fill it with positive thoughts, your life will start to change"

Understanding how trauma changes neural pathways is crucial for recovery. Biochemical addiction plays a major role in abusive relationships marked by trauma bonding. Neurochemicals such as oxytocin, serotonin, dopamine, cortisol, and adrenaline reinforce the victim-perpetrator attachment, making it hard to break away. The amygdala controls emotional responses to trauma by releasing stress hormones that activate the sympathetic nervous system, leading to symptoms like sweating, trembling, and anxiety. Chronic fight-or-flight responses from ongoing stress can cause long-term emotional and physiological issues. People raised by angry caregivers often become hypervigilant and suppress their own needs for self-protection, which may result in feelings of isolation, shame, guilt, and chronic anxiety as adults. Encountering anger can trigger old fears, making vulnerability and emotional connection difficult, leading to psychosomatic problems.

Unresolved trauma keeps the nervous system in a state of chronic imbalance, making it hard to recover even after the stressor is gone. Symptoms of a dysregulated nervous system include sleep issues; difficulty regulating emotions; chronic or long-term physical pain; sensory sensitivity; gut issues; brain fog; memory issues; hyperactivity and/or frequent dissociation; and feel wired and tired. It is possible to move the nervous system from a dysregulated state through mindful movement practices that emphasise body awareness and the mind-body connection, such as somatic exercises. As stress, trauma, and tension are gradually released from the body, individuals may experience greater calmness both physically and emotionally, along with a reduction in pain. This process can lead to increased energy, improved sleep patterns, and enhanced ability to focus on relationships and professional responsibilities. With these observed improvements, individuals can work towards effective self-regulation of the nervous system, providing an opportunity for significant overall wellbeing.

Unhealthy power dynamics can undermine relationships and negatively impact mental health. Engaging with a qualified therapist can assist in addressing the complex issues that may arise, such as those resulting from love bombing. Individuals affected may encounter a range of emotions, including depression, anxiety, anger, sadness, confusion, grief, and feelings of loss. As a result of manipulative tactics—particularly those associated with narcissistic behaviour—victims may also exhibit physical symptoms known as Identifiable Victim Markers (IVMs). Symptoms include:

Anxiety: Persistent manipulation and criticism may contribute to ongoing anxiety.

Argumentative: Individuals may exhibit argumentative behaviour as a means of self-defence.

Avoidance: Avoidance of certain people or situations that could lead to conflict.

Compliance: Individuals may comply with demands to prevent disagreement.

Conditioned Beliefs: Development of beliefs consistent with external influences.

Defensiveness: A tendency to defend oneself regularly in response to frequent criticism.

Detachment: Emotional distancing to manage distress.

Denial: Downplaying or denying the presence of problematic behaviour as a coping mechanism.

Dismissive: Less consideration given to one's own feelings or needs as an adaptive response.

Isolation: Reduction in contact with friends and family members who might notice changes.

Labelling: Acceptance of negative labels from others.

Minimisation: Downplaying the effects or severity of challenging experiences.

Nervousness: Ongoing nervousness associated with stressful environments.

Non-accountable: Difficulties with accepting responsibility, possibly in relation to repeated blame.

Reactive: Increased emotional responses to triggers related to past experiences.

Reactive narcissism: Adopting narcissistic traits in reaction to similar behaviours.

Self-blame: Assigning personal responsibility for negative events or outcomes.

Self-criticism: Frequent self-critical thoughts, sometimes reflecting prior feedback from others.

Self-doubt: Uncertainty about one's own abilities, value, or judgement.

Narcissistic abuse can result in substantial, long-term emotional and psychological harm, often manifesting as feelings of worthlessness, isolation, confusion, and self-doubt. Individuals affected may also experience anxiety and depression. This form of abuse typically constitutes a deliberate and manipulative behavioural pattern in which the narcissist seeks to exploit and control the victim. Effective recovery strategies include recognising the abuse, establishing clear boundaries, understanding its impact on self-worth, and obtaining appropriate support. Victims have multiple avenues available to initiate their recovery process:

Therapeutic support: Counselling and therapy can be vital in healing from narcissistic abuse.

Support groups: Finding support from others who have experienced similar abuse.

Education and awareness: Understanding narcissistic behaviours can help in recognising and combating abuse.

Legal Protection: In some cases, legal measures might be necessary to protect oneself.

Identifying strategies for fostering positive future relationships can contribute to ongoing wellbeing. Learning about the signs of narcissistic abuse can help individuals recognise and respond appropriately to such behaviour. Establishing and maintaining clear boundaries during interactions with individuals who may present a risk of abusive behaviour can also support the identification of potential warning signs.

Chapter Thirty-Four

ANXIETY

"Every positive thought is an investment into the life you're building"
Donna Siggers

Anxiety is a common psychological response that involves cognitive, emotional, and physiological changes when an individual encounters threats, stress, or pressure. It often manifests as worry, tension, or fear, triggered by perceived dangers or certain situations. While anxiety is a regular part of human experience, it is classified as a mental health concern if it interferes with an individual's ability to function effectively. Anxiety may be intense or persistent, and fears or worries can sometimes be disproportionate to the actual circumstances, potentially leading to avoidance behaviours. Persistent distress can also result in panic attacks and difficulties in performing activities or daily tasks. Anxiety is commonly observed among individuals who have experienced narcissistic abuse. Survivors may develop anxiety disorders due to the manipulative, controlling, and invalidating behaviours characterising such interactions. These individuals frequently report persistent nervousness and apprehension, especially in response to the abuser's unpredictable conduct.

Although occasional anxiety is considered a normal reaction to daily stressors, those with anxiety disorders experience intense, excessive, and prolonged worry or fear that affects routine activities.

Such disorders often manifest as recurrent episodes of acute anxiety, fear, or panic—commonly known as panic attacks—which typically arise rapidly. These episodes can be difficult to manage and may be disproportionate to the perceived or actual threat, sometimes continuing for extended durations.

There are many symptoms of anxiety: Feeling nervous, restless, or tense; having a sense of impending danger, panic, or doom; having an increased heart rate; breathing rapidly (hyperventilation); sweating; trembling; feeling weak or tired; trouble concentrating or thinking about anything other than the present worry; having trouble sleeping; experiencing gastrointestinal (GI) problems; and having the urge to avoid situations or objects that trigger anxiety. Types of anxiety include:

Agoraphobia: Agoraphobia is characterised by a persistent fear and avoidance of environments or situations where an individual might experience panic, or feelings of being trapped, helpless, or embarrassed.

Anxiety disorders: Anxiety disorders caused by medical conditions involve symptoms of anxiety or panic that are directly linked to a physical health issue.

Generalised anxiety disorder: Generalised anxiety disorders are characterised by ongoing and excessive anxiety and worry about various activities or events, including routine daily tasks. The level of worry does not correspond to actual circumstances, is challenging to manage, and can affect physical wellbeing. This condition may also present with other anxiety disorders or depression.

Panic disorder: Panic disorders are marked by recurrent, sudden episodes of intense anxiety or fear, known as panic attacks, which peak within minutes. Symptoms include a sense of impending doom, shortness of breath, chest pain, or palpitations. These attacks can cause ongoing worry or avoidance of similar situations.

Selective mutism: Selective mutism refers to a condition in which children consistently do not speak in certain environments, such as school, while communicating normally with family and friends in other settings. This can impact their learning and social interactions.

Separation anxiety disorder: Separation anxiety disorder is a paediatric condition marked by pronounced anxiety that is disproportionate to the child's developmental stage. This disorder typically arises in response to separation from individuals who fulfil parental or caregiving roles.

Social anxiety disorder (social phobia): Social anxiety disorder, also called social phobia, is characterised by significant anxiety, fear, and avoidance of social situations arising from concerns about embarrassment, self-consciousness, or being negatively evaluated by others.

Specific phobias: Specific phobias involve significant anxiety when an individual encounters a particular situation and may result in avoidance behaviour. In some cases, phobias can trigger panic attacks.

Substance-induced anxiety disorder: Substance-induced anxiety disorder is defined by pronounced symptoms of anxiety or panic that arise directly from substance misuse, including the use of drugs or alcohol, medication administration, exposure to toxic substances, or withdrawal.

Other specific anxiety disorder: Other specific anxiety disorder and unspecified anxiety disorder are classifications for types of anxiety or phobias that do not meet the criteria for other identified anxiety disorders but can influence daily functioning.

Chapter Thirty-Five

DEPRESSION

---

*"A big part of depression is feeling really lonely,
even if you're in a room full of a million people"*
Lilly Singh

Depression is a mood disorder defined by persistent feelings of sadness and a decreased interest in previously enjoyed activities. Commonly termed major depressive disorder or clinical depression, it affects both emotional regulation and cognitive processes, often resulting in a range of physical and psychological complications. Affected individuals may have difficulty performing daily tasks or maintaining engagement in once pleasurable pursuits. Unlike transient mood changes, depression typically requires professional intervention and sustained treatment. Research demonstrates that most patients experience symptom relief with pharmacological therapy, psychotherapy, or a combination thereof. The clinical presentation of depression varies among individuals, encompassing symptoms such as low mood, hopelessness, reduced motivation, and episodes of tearfulness; anxiety is also frequently observed alongside depressive symptoms. In addition to its psychological impact, depression can produce physical manifestations, including chronic fatigue, disturbed sleep patterns, diminished appetite, reduced libido, and somatic pain. Depression is a condition that may present across all age groups. The following outlines the symptoms that present within differing age ranges:

Depression symptoms in young children: Sadness; irritability; clinginess; worry; aches and pains; refusing to go to school; and being underweight.

Depression symptoms in teenagers: Sadness; irritability; feeling negative and worthless; anger; poor performance or poor attendance at school; feeling misunderstood and extremely sensitive; using recreational drugs or alcohol; eating or sleeping too much; self-harm; loss of interest in normal activities; and avoidance of social interaction.

Depression symptoms in adults:

> *Psychological symptoms:* continuous low mood or sadness; feelings of hopelessness; low self-esteem; feeling tearful; feeling guilt-ridden; feeling irritable and intolerant of others; lack motivation or interest in things; finding it difficult to make decisions; not getting any enjoyment out of life; feeling anxious or worried; having suicidal thoughts or thoughts of harming oneself.

> *Physical symptoms:* Moving or speaking more slowly than usual; changes in appetite or weight; constipation; unexplained aches and pains; lack of energy; low sex drive; disturbed sleep (finding it difficult to fall asleep at night or waking up very early in the morning).

> *Social symptoms:* Avoiding contact with friends and taking part in fewer social activities; neglecting hobbies and interests; and having difficulties in home, work, or family life.

The severity of depressive symptoms varies:

Mild depression may involve a consistent low mood, and have some impact on daily life.

Moderate depression can substantially affect an individual's daily functioning. It is advised to use a combination of psychotherapy and antidepressant medication when treating moderate depression.

Severe depression can cause extreme challenges for individuals to manage everyday activities, and in some cases may present with psychotic symptoms. Patients with severe depression are typically referred to specialist mental health teams for intensive psychotherapy and pharmacological interventions. Severe depression can lead to thoughts of suicide and feelings that life is not worthwhile. Early consultation with a healthcare professional is recommended for individuals experiencing symptoms, as early intervention can support recovery.

Clinical depression presents differently across individuals and may last for several weeks or months, often affecting work, social interactions, and family life. Treatment typically includes lifestyle modifications, medication, and psychological therapies.

Certain events, such as bereavement, loss of employment, or childbirth, may trigger depression. A family history of depression increases an individual's risk, although depression can also occur without clear cause. Treatment options include lifestyle modifications, talking therapies, and medication, with recommendations depending on symptom severity. There are many circumstances that might result in depression, including:

Postnatal depression may affect new mothers, fathers, or partners after the birth of a child. Healthcare professionals often recommend therapeutic interventions, such as counselling and pharmacological treatments, as potential management strategies. Symptoms might present as: Being down, upset, or tearful; restless, agitated, or irritable; guilty, worthless, and self-critical; empty and numb; isolated and unable to connect with others; a lack of enjoyment in life or previously liked activities; detached from reality or a sense that things are not real; lacking in self-confidence or self-esteem; hopeless and despairing; hostile or indifferent towards partner; hostile or indifferent towards the baby; or suicidal thoughts. Additionally, there might also be loss of concentration; struggles sleeping, even when the opportunity arises; reduced appetite; and low libido.

Bipolar disorder, sometimes known as *manic depression*, is distinguished by recurrent episodes of both depressive and elevated (manic) mood states. Depressive symptoms typically mirror those seen in clinical depression, whereas manic episodes may manifest as increased impulsivity, including behaviours such as gambling, excessive spending, and participation in high-risk activities.

Seasonal affective disorder (SAD), also called *winter depression* typically occurs during the winter months.

Many individuals experiencing depression benefit from implementing lifestyle modifications such as increasing physical activity, reducing alcohol consumption, stopping smoking, and maintaining a nutritious diet. Engaging with self-help literature or participating in support groups can also be valuable, as these approaches facilitate a deeper understanding of depressive symptoms and their potential causes. Sharing experiences with peers in similar circumstances often fosters a greater sense of support and validation.

Depression frequently develops gradually, making it challenging to recognise at its onset. Many attempt to manage symptoms independently without realising they are unwell, and it is sometimes a friend or family member who first identifies changes in behaviour. Differentiating between grief and depression can be complex, as both share overlapping symptoms; however, key distinctions exist. Grief is a natural response to loss, whereas depression constitutes a medical condition. While those grieving typically experience intermittent sadness but retain the ability to find enjoyment and hope for the future, individuals with depression may withdraw from daily activities for extended periods.

There is no single cause of depression; it may arise due to various factors and can be triggered by numerous events. For some, stressful or distressing life circumstances—including bereavement, divorce, illness, redundancy, or financial difficulties—may contribute to its onset. Multiple contributing factors frequently intersect, precipitating depression. For instance, an individual may begin to feel low following illness and subsequently encounter a traumatic event, such as bereavement, which exacerbates depressive symptoms. People often describe a downward spiral where one adverse event leads to further isolation and unhealthy coping mechanisms, thereby intensifying depressive symptoms.

The risk of depression rises in the face of stressful events, particularly when social connections deteriorate and individuals become isolated. Certain personality characteristics, such as low self-esteem or excessive self-criticism, may increase susceptibility to depression. Such vulnerability might be attributed to genetic predispositions, early life experiences, or both. Isolation from family and resulting loneliness can also serve as significant contributing factors.

Initiative-taking steps can assist in alleviating depressive symptoms and supporting recovery. Adhering to prescribed medication regimes is essential, even during periods of improvement, as premature discontinuation may result in relapse. Consulting healthcare professionals regarding any questions or concerns about treatment is highly recommended, and informational leaflets provided with medications offer additional guidance. Regular exercise combined with a balanced diet can expedite recovery and promote overall health. Nutritious eating habits contribute to improved mood and general wellbeing, while physical activity can elevate mood, reduce stress and anxiety, stimulate the release of endorphins, and enhance self-esteem. Exercise also provides distraction from negative thoughts and opportunities for increased social interaction. In today's fast-paced environment, individuals may inadvertently overlook their surroundings. Practicing mindfulness allows for greater awareness of the present moment and facilitates self-understanding. Mindfulness involves recognising both internal sensations and external stimuli as they occur, which can help disengage from unhelpful thought patterns and improve emotional regulation. This practice supports the redirection of negative thoughts and feelings into more constructive channels, enhancing the capacity to address challenges effectively.

Caregivers supporting individuals with depression should recognise the broader impact on themselves and others involved. It is important to acknowledge these challenges and seek appropriate resources and support to navigate the caregiving process productively. The following outlines intervention steps that consider both the person with depression and the caregiver:

Listen without judgement: Establish an environment where individuals experiencing depression can express their feelings openly, free from criticism or unsolicited advice.

Validate feelings: Recognise that all emotions are legitimate and that it is acceptable to experience periods of discomfort.

Care: Reassure the individual experiencing depression that they are valued and supported; their condition does not alter the respect and care held for them.

Be patient: Recognise that recovery from depression is a gradual process, characterised by periods of progress as well as occasional setbacks.

Suggest therapy: Encourage the individual to consider seeking assistance from a qualified mental health professional, such as a therapist or counsellor.

Help find resources: Facilitate the process of identifying a therapist, support group, or other appropriate mental health services that best meet their individual needs.

Encourage medication adherence: If medication is prescribed, assist with adhering to the dosage instructions and ensure all scheduled medical appointments are attended.

Promote physical activity: Recommend light physical activity, such as walking, as it may contribute positively to emotional wellbeing.

Encourage a balanced diet: Suggest healthy eating habits and discourage unhealthy coping mechanisms such as alcohol or drug use.

Promote social interaction: Encourage individuals to interact with friends and family, or to participate in social activities, even if they may not initially feel inclined to do so.

Important considerations:

> *Safety:* If there are concerns about the individual's safety or indications of suicidal thoughts, encourage seeking prompt professional assistance.

> *Individualised Approach:* Approaches may vary in effectiveness depending on the individual's specific

needs and preferences; personalised strategies can yield better results.

Stigma: Reducing the stigma associated with mental health and promoting open dialogue about depression is essential.

Set boundaries: When supporting an individual experiencing depression, it is essential for those around them to safeguard their own mental health, too.

Seek support: Consider reaching out to friends, family, or a mental health professional if supporting someone with depression feels challenging.

Antidepressants are medications prescribed to alleviate the symptoms of depression. They function by increasing the levels of neurotransmitters in the brain, which are chemical messengers that facilitate communication between nerve cells and play a critical role in various physiological processes. Specifically, serotonin and noradrenaline are neurotransmitters associated with mood regulation and emotional wellbeing. There are two main types of antidepressants: Selective Serotonin Reuptake Inhibitor (SSRIs) and Tricyclic Antidepressants (TCAs). While their efficiency is comparable, significant differences exist regarding patient suitability and potential side effects. Selective serotonin reuptake inhibitors are typically the first line of treatment, as their side effect profiles are often more manageable and they present a lower risk of severe complications in overdose cases. It is essential for patients to adhere closely to both their physician's guidance and the instructions provided in the accompanying medication leaflet. Additionally, promptly reporting any unexpected or concerning reactions to a healthcare professional is advised.

Chapter Thirty-Six

FEAR

"Only dread one day at a time"
Charlie Brown

Fear functions as an essential survival response sought by some individuals through experiences such as roller-coaster rides or engagement with horror media, while others endeavour to avoid it. As a psychological experience, fear elicits a physical reaction within the body: When detected, the amygdala activates the nervous system, initiating the fear response. This process releases stress hormones such as cortisol and adrenaline, resulting in elevated blood pressure, increased heart rate, and accelerated breathing. Blood flow is redirected from the heart to the limbs, facilitating potential survival actions, which may include fighting, fleeing, or freezing. Collectively, these reactions constitute the fight, flight, or freeze response. Concurrently, neural activity shifts; certain brain regions downregulate as others prepare for action. The amygdala's involvement impairs the cerebral cortex—the area responsible for reasoning and judgement—thereby reducing capacity for clear decision-making. Subsequent behaviour may involve vocalisations or other instinctive reactions that occur regardless of the legitimacy of the threat. Fear may persist beyond the resolution of actual or perceived threats due to excitation transfer whereby heightened arousal remains post-experience. This phenomenon is prevalent after watching simulated fear scenarios, such as viewing films, when the dopamine production continues to make a person 'jumpy'.

Distinguishing fear from phobia is crucial. Whereas fear is a common, adaptive emotional response to specific stimuli or events, phobias are characterised by sustained impairment in daily functioning and marked distress. Extreme avoidance behaviours, such as staying away from water, spiders, or people, suggest the presence of a phobic disorder. Fear is a universal biological phenomenon that promotes safety and survival, though excessive experience can yield negative outcomes.

Contemporary society presents myriad risks, contributing to prevalent fears that affect health and wellbeing. These anxieties derive partly from advancements following World War II, including industrial and technological developments that have introduced both benefits and hazards. Improvements such as plastics, pesticides, nuclear power, mobile technology, biotechnology, and global mobility have contributed to increased life expectancy, reduced infant mortality, and control of major diseases through vaccination. Water and air quality have improved, suggesting increased public health and safety compared to previous eras. Nonetheless, progress incurs costs: Environmental changes, population growth, disease transmission due to enhanced transportation, and global availability of food have led to issues such as obesity and cardiovascular disease. Individual technologies and projects—including chlorofluorocarbons, mobile phones, and nuclear energy—present new risks, perpetuating pervasive concern. Simultaneously, the current era provides unprecedented immediacy and accessibility of information. Enhanced communication networks and widespread internet access allow for rapid dissemination of knowledge, including notifications of potential risks. This phenomenon has produced continuous streams of dramatic and alarming news coverage, amplified perceptions of global danger and has contributed to widespread anxiety.

PART SIX

BEGINNING TO HEAL

"Healing of the soul first; then healing of the mind and body will follow"
Zhi Gang Sha

Individuals may encounter a broad spectrum of emotions throughout the recovery process following abuse. The trajectory of healing varies significantly between individuals, and there is no universally applicable approach to recovery. All emotional responses during this period are recognised as valid. Fluctuations in mood—ranging from confidence and positivity to anxiety and sadness—are commonly observed.

Controlling behaviours are frequently observed in abusive relationships, which may hinder an individual's transition to independence. Many people report difficulty in making autonomous decisions following such experiences. It is common for survivors of abuse to experience feelings of longing for their previous partners or to contemplate renewing contact. Additionally, survivors may present with symptoms such as nightmares and flashbacks, which can be indicative of post-traumatic stress disorder (PTSD). Ongoing fear or heightened vigilance is also possible. Although recovery from emotional or physical abuse can be complex and time-consuming, prioritising therapeutic progress can gradually mitigate these challenges. Exiting a relationship involving trauma bonding can present various challenges and complexities. The process may involve significant emotional responses. Identifying the presence of a trauma bond is an important step toward addressing patterns of abuse and reducing their likelihood in future relationships. Identifying the existence of a trauma bond and acknowledging related experiences and emotions can clarify that abusive circumstances are not attributable to individual fault. This recognition helps validate feelings and challenges, supporting informed decision-making. Trauma bonds often exist where denial and normalisation are common. Recognising a trauma bond directly addresses tendencies to downplay or overlook abusive behaviours and their impact on wellbeing. Such acknowledgment is important for correcting misconceptions that abuse is acceptable or typical. Once recognised, it becomes possible to examine the patterns and dynamics within the relationship. This awareness enables the identification of warning signs and manipulative tactics, potentially preventing similar patterns in future relationships. Identifying a trauma bond may support individuals in regaining control over their lives, facilitate more objective evaluations of the relationship, and inform decisions regarding personal safety and recovery. Breaking a trauma bond involves recognising and addressing relevant dynamics and patterns, which can disrupt cycles of abuse and influence future interactions. This approach can help establish boundaries and self-care practices. Accessing professional counselling and guidance is advisable when preparing to leave an abusive relationship. Trauma therapy professionals have expertise in trauma bonding and the effects of abuse, and can offer with safety planning, risk management, and transition out of the situation. Additionally, they provide psychoeducation to enhance understanding of trauma bonding and its psychological and emotional impacts and offer a supportive environment for expression without judgment. With professional support, individuals can be better prepared to exit abusive relationships and begin the process of healing and recovery.

Restoring independence and personal identity is a critical component of overcoming an abusive relationship. Engaging in purposeful activities that align with one's values and long-term objectives contributes significantly to rebuilding self-confidence and autonomy. By investing time in individual interests, aspirations, hobbies, support networks, or professional therapy, survivors of abuse can identify personal strengths and cultivate a resilient sense of self. Regaining independence equips individuals to make informed decisions that promote overall health and well-being.

In circumstances involving abuse, meticulous documentation of incidents is highly recommended. Where it is safe and feasible to do so, maintain detailed records of each occurrence, including dates and relevant specifics, to substantiate personal accounts. This documentation may prove invaluable should legal proceedings or protective actions become necessary. Additionally, preserving physical evidence—such as text messages, emails, or photographs of injuries—can provide compelling support for any claims made. Throughout this process, prioritising personal safety is essential; it is advisable to seek assistance from professionals or organisations specialising in domestic violence to ensure appropriate measures are taken to safeguard wellbeing. Abusive relationships pose substantial risks to both physical and emotional health. Making the decision to exit such a situation is a crucial measure in safeguarding oneself against continued harm, as prolonged abuse can lead to severe psychological and emotional consequences. For a secure departure, it is advisable to formulate a detailed safety plan with the guidance of professionals experienced in domestic violence intervention. Involving trusted friends, family members, or established support networks who are informed of the plan to leave can offer additional assistance. Moreover, seeking counsel from legal professionals specialising in domestic violence and family law can provide critical insights regarding individual rights and the spectrum of legal protections available.

Trauma bonding extends beyond romantic partnerships and can also occur within friendships or family relationships. These adverse dynamics are characterised by unhealthy interactions rooted in shared traumatic experiences, emotional dependence, and maladaptive coping mechanisms. Individuals who have endured significant or distressing events together may develop strong attachments, often seeking support and validation from each other. This dynamic can lead to an overreliance on the relationship for emotional fulfilment, thereby reinforcing unhelpful behaviours as coping strategies. Additionally, such relationships may emphasise validation at the expense of establishing healthy boundaries and accountability.

Chapter Thirty-Seven

SETTING HEALTHY BOUNDARIES

"Do not give your past the power to define your future"
Dhiren Prajapot

Individuals who lack clearly defined boundaries are at increased risk of exploitation by those displaying narcissistic characteristics, which may lead to the depletion of energy, time, financial resources, or other personal assets. The establishment of appropriate boundaries is an essential aspect of self-care and accountability. Prioritising personal wellbeing serves to improve interpersonal relationships and supports a more fulfilling life. Setting boundaries with family members and close associates can reduce potential adverse effects on mental health and emotional stability. Persistent exposure to negativity by those demonstrating narcissistic tendencies can influence behaviours within social circles. While emotionally or physically distancing oneself from such people may be challenging, it remains advantageous for evaluating and strengthening personal relationships. Certain questions may be posed during the process of establishing healthy boundaries, which can assist survivors of narcissistic abuse in evaluating whether prospective relationships are likely to be healthier: Does this person cause pessimistic feelings; do conversations with someone close cause a feeling of being emotionally drained; does the person demand immense amounts of time; does this person always expect others to pay in social settings; and does this person use manipulative behaviour to convince others to attend events they don't wish to. Affirmative responses suggest that establishing boundaries and, when necessary, declining requests are vital for safeguarding wellbeing. Although setting boundaries may initially be challenging—especially for those unfamiliar with this practice—approaching it as a form of self-compassion intended to promote constructive interactions can facilitate a smoother adjustment. By prioritising personal needs, individuals are better equipped to efficiently manage their time, resources, emotional wellbeing, physical health, mental acuity, and relationships.

Chapter Thirty-Eight

HEALING FROM NARCISSISTIC RELATIONSHIPS

"Nobody can save you but yourself"

For a relationship with a narcissistic partner to continue, it is necessary that they are open to change and acknowledge the importance of developing appropriate and healthy alternatives to problematic behaviours. Providing an opportunity for them to learn and adjust is an initial step towards these adjustments. In some cases, discussions about healthy boundaries, if previously held, may need to be revisited. If these boundaries continue to be disregarded or behaviour deteriorates further, it may be necessary to consider ending the relationship in a safe way. Such situations can be complex and challenging. Often, individuals may find it difficult to trust new partners, or their own judgment, in future relationships. Nonetheless, it is possible to move forward following these experiences by being transparent about past relationships and establishing healthy boundaries at the start of new ones.

Chapter Thirty-Nine

COMMUNICATING WITH A NARCISSISTIC PERSON

"A narcissist doesn't see other's perceptions, they don't do compromise, they only see it as you starting an argument"

Engaging in communication with people who display narcissistic traits can present unique challenges. These individuals may misinterpret comments, express uncertainty about the speaker's intentions, or steer discussions toward their own interests. Consequently, they may shift responsibility onto the other party, suggesting that the individual is being unreasonable, lacks understanding of their viewpoint, or is provoking unnecessary debate. Remaining calm and using a respectful tone is less likely to provoke language, belittling remarks, or retaliatory responses. Whereas, challenging, correcting, or embarrassing an abusive person may increase the likelihood of arguments or resistance. It is beneficial to use 'I' statements as this is a particularly effective way to promote constructive communication. They convey one's needs, perspectives, and emotions, while concentrating on personal experience rather than emphasising the other person's shortcomings, facilitating more positive dialogue. For example, instead of stating, 'you never listen to me' one might express, 'I feel that I was not heard earlier'. This method can foster greater empathy and productivity during interactions. Individuals who exhibit narcissistic characteristics may respond more positively to the following conversation openers: I feel, I hear, I want, and I wish.

Advocating wellbeing while interacting with individuals who may exhibit narcissistic traits can, sometimes, be difficult. Sharing perspectives or concerns may appear confrontational to them. Nevertheless, clear communication and self-assertion are important. Setting boundaries and restating key points, when necessary, can help maintain clarity, even if the other party responds with criticism or attempts to redirect the conversation. Despite this, it remains critical that clear boundaries are established. Positive communications might be: I will no longer allow you to speak to me this way; I refuse to engage in this conversation while you're yelling; let's talk and I'll stay; I won't be continuing this conversation if you insult me or belittle my feelings. Engaging in productive dialogue is achievable. Although it may be challenging at times, consider keeping these tips in mind the next time conversation arises: Stay calm and respectful; share feelings with 'I' statements; preserve principles without being defensive; maintain boundaries, but try to be empathetic; and rely on a support system.

Deflecting blame onto others, regardless of circumstance, is a common trait of narcissism. Remaining calm and refraining from accepting accountability for matters beyond your control is advised. Narcissistic behaviours are frequently observed in the context of abusive relationships, and efforts to modify such behaviours are typically challenging and unlikely to succeed unless initiated by the individual concerned. Engaging with those displaying narcissistic characteristics can present considerable difficulties and lead to emotional distress. When navigating these dynamics, it is beneficial to seek advice from trusted colleagues, family members, or other reliable support networks. Building and maintaining connections with positive and supportive individuals contributes to overall wellbeing and offers stability. Consulting an experienced mental health professional is recommended; they can provide evidence-based coping strategies and tailored guidance for managing interactions with persons who exhibit narcissistic personality traits. Implementing coping mechanisms, such as taking breaks when triggers are identified, can facilitate more effective engagement and allow time for reflection and composure. Should these patterns persist and negatively impact mental health, it is important to evaluate the relationship. Ending a relationship, particularly with a long-term partner or close family member, is a complex decision that should be made with careful consideration. It is important to be able to identify when it is time to walk away from someone with narcissistic tendencies. Their behaviour might convey feelings of: Feeling threatened; feeling emotionally, verbally, or mentally abused; being repeatedly disrespected; of being manipulated; being controlled; and that conversations are consistently escalating or becoming heated.

Chapter Forty

WHY IS IT HARD TO LEAVE

"I didn't leave because I stopped loving you. I left because the longer I stayed, the less I loved myself"

Exiting narcissistic relationships is often a complex and challenging process. Victims are frequently questioned about why they did not leave earlier—which they may also ask themselves while attempting to understand their experiences. The situation is rarely straightforward. Narcissistic abuse can severely damage an individual's sense of identity and self-worth. Perpetrators of such abuse do more than gaslight; they systematically erode the victim's trust in their own memories, instincts, and value. By the time an individual considers leaving, they may genuinely believe they are at fault—an outcome of sustained psychological manipulation. Narcissists often invest significant effort in understanding their victims, initially displaying behaviours that align with their deepest needs and presenting themselves as ideal partners. Leaving such a relationship entails mourning the perceived reality of that partnership. The narcissist strategically exploits the victim's empathy, loyalty, and compassion, alternating between displays of remorse or distress and subsequent acts of emotional harm. Each episode of remaining in the relationship tends to deepen the psychological impact.

Departing from abusive relationships is seldom a discreet process, as narcissists may retaliate through tactics such as love bombing, making threats, exerting financial control, employing manipulative behaviour, and enlisting mutual acquaintances. Their primary objective is to reassert control, often by distorting events and influencing others to adopt their perspective. This can result in the victim being portrayed as unstable, irrational, or disloyal, thereby further isolating them. Narcissists typically lack insight into their own actions, as acknowledging fault contradicts their constructed narrative.

Individuals subjected to trauma bonding within these dynamics endure repeated cycles of harm followed by comfort, a process that is both subtle and deeply damaging. Publicly, narcissists may present as composed and affectionate, while privately engaging in patterns of cruelty. Victims of this form of abuse do not experience genuine affection; rather, their attachment resembles an addiction, making the process of leaving analogous to withdrawal.

Chapter Forty-One

THE ORIGIN OF FEELING

A foundational aspect of recovery from an abusive relationship involves reflecting on the last instance when one experienced pleasure, even if it was as subtle as listening to birdsong or watching a sunset. It is also beneficial to acknowledge an unpleasant experience, such as illness or interpersonal conflict. This practice helps to clearly differentiate between pleasure and displeasure. These sensations represent the brain's continuous interpretation of physiological signals originating from internal organs, tissues, circulating hormones, and the immune system. The body's dynamic functions—including cardiac, pulmonary, and digestive activities— produce a range of basic affective states, spanning pleasant to unpleasant and calm to agitated, while also encompassing neutral experiences. These foundational emotional states play an important role in shaping more complex emotions, such as joy and sadness, and are essential for understanding the relationship between emotional experiences and overall physical and psychological health.

The mind functions through the integrated interaction of the brain and body. The brain perpetually updates its predictions using prior knowledge, refining these anticipations in response to specific contextual requirements. This process is demonstrated when individuals misidentify a stranger as someone familiar or notice altered sensations after moving from a boat to stable ground. By processing sensory information and adapting perceptions accordingly, individuals can experience new situations rather than routine repetition. Neural processes, defined by ongoing prediction and adjustment, form the basis of perception and play a central role in learning. They also play a critical role in enabling adaptive responses to environmental stimuli. For instance, the act of catching a ball demand more than simple visual tracking; predictive processing is necessary to estimate speed, trajectory, and the probability of successful interception. The brain generates anticipatory simulations prior to the receipt of sensory information, synthesising visual, auditory, and other data to optimise preparedness for action. Effective performance occurs when these predictions closely correspond with actual sensory input, facilitating coordinated movement. In contrast, significant deviations between internal forecasts and external events can lead to errors, thereby diminishing the likelihood of success. Accurate predictions are therefore fundamental to effective response execution. The dynamic interaction between prediction and error determines the extent to which experience is influenced by external stimuli versus internal constructs. Exclusive reliance on predictive processes, without reference to sensory input, can result in subjective experiences that are disconnected from objective reality. Conversely, accurate predictions facilitate coherent perception and intentional action by ascribing meaning to sensory information. This predictive framework is considered a foundational aspect of brain function, oriented not merely toward anticipating future events, but toward actively simulating them.

Chapter Forty-Two

THE STAGES OF GRIEF AND LOSS

"There is no normal way to grieve. Except for how we each do it"

Developing patience can support recovery from narcissistic relationships and assist in building new ones. Individuals who have experienced abuse may be cautious when starting a new relationship. Recognising that there are many engaging experiences beyond past relationships, and allowing emotions to arise, can contribute to a positive future. Challenging thoughts and feelings may occur during this adjustment, which is a normal part of breaking connections and forming new ones. It is important for both partners to maintain their own identities, establish healthy boundaries, uphold a clear sense of reality, and share power within the relationship constructively. Reengaging in activities previously limited by a past partner can also help individuals reclaim their sense of self, develop healthier relationships, and foster personal growth.

Grief is a natural aspect of the recovery process and may be overwhelming at times. Developing an understanding of grief facilitates recognition of one's emotions and responses, which can significantly support healing and the formation of new relationships. Grief and loss are commonly conceptualised as processes encompassing denial, anger, bargaining, depression, and acceptance. These stages are not inherently linear or universal, but are, instead, an individual experience that is distinct and personal. Individuals may revisit certain phases throughout their journey. Grief extends beyond feelings of sadness or longing; it is a complex, multifaceted process characterised by fluctuations over time and can span several months or years. Developing an understanding of this process may facilitate greater insight into emotional responses and coping mechanisms. The stages of grief and loss are:

Denial: Denial refers to the psychological response in which an individual convinces themselves that no change or loss has occurred, maintaining the perception that circumstances remain unchanged. Frequently identified as the initial stage of the grieving process, denial serves as an emotional safeguard against the immediate impact of significant loss or trauma. This coping mechanism provides individuals with temporary distance from reality, allowing for a gradual adjustment to new circumstances. A typical manifestation of denial includes persisting with regular daily routines as though nothing has changed, effectively creating a buffer between the shock of loss and acceptance of reality, thereby facilitating adaptation at a manageable pace.

Anger: Anger represents a natural coping mechanism that can emerge when individuals attribute blame to themselves or others following a loss. This response often serves as an attempt to rationalise circumstances that may appear arbitrary or unjust. Frequently, anger conceals underlying pain and vulnerability. Individuals should not experience guilt or shame for feeling angry, as this emotion constitutes a normal and healthy means of processing accumulated feelings. Over time, anger typically diminishes as alternative coping strategies develop.

Bargaining: Bargaining is a common psychological response in which individuals attempt to reconsider past events to influence their outcomes. Many people derive comfort from the notion that alternative actions might have produced different results, suggesting that circumstances were not entirely beyond their control. This stage frequently involves internal negotiation as a means of coping with grief. Bargaining may serve as an effort to restore a sense of control or hope when facing emotionally challenging situations. Additionally, seeking alternative solutions can foster openness to diverse perspectives and innovative approaches.

Depression: Depression is, usually, the stage in which the reality of loss becomes clear. During this period, people often perceive their emotions as more concrete and lasting. This phase indicates the full emotional response to loss and may be more intense than other stages of grief. While this timing might not align with expectations of feeling better, the grieving process tends to be complex. Depression related to grief differs from clinical depression in that it can fluctuate throughout the day, sometimes fading entirely before returning unexpectedly. In contrast, clinical depression usually presents with persistent symptoms over time.

Acceptance: Accepting loss does not entail forgetting the individual or circumstances involved. Rather, acceptance occurs when an individual acknowledges death, trauma, or other events, and begins to formulate a clear perspective for the future. It involves moving forward and continuing life alongside the experience of loss, while still honouring the person at its core. Individuals in the acceptance phase often demonstrate increased calmness and reflection compared to those in earlier stages of grief. Acceptance is about reconciling with the events that have transpired and determining appropriate steps for the path ahead.

An effect of grief and loss can be guilt. Guilt is defined as an emotional response characterised by remorse over previous actions, typically arising when an individual perceives that harm has been caused or that personal moral standards have not been upheld. Individuals may respond to situations differently; what one person perceives as significant, another may not. Experiencing guilt is a common aspect of human behaviour and can occur in response to various circumstances. There are two categories of guilt:

Appropriate guilt, which can cause uncomfortable feelings and plays a critical role in regulating social behaviour. Legitimate feelings of guilt indicate that one's conscience and cognitive functions are operating effectively, preventing the repetition of errors. This process facilitates learning from past actions and encourages future behavioural adjustments. Individuals who exhibit a tendency towards guilt proneness are thought to demonstrate a heightened awareness of both their own emotions, and those of others.

Irrational guilt: Taking undue responsibility for situations or overestimating the impact of one's actions can be detrimental if not appropriately addressed. Excessive irrational guilt has been associated with mental health conditions such as anxiety, depression, dysphoria, and obsessive-compulsive disorder (OCD). Individuals experiencing irrational guilt may perceive themselves as burdensome to their loved ones and those around them. This can lead to diminished concentration and productivity, lowered mood, heightened stress, and sleep disturbances. Consequently, personal relationships, daily functioning, and overall outlook on life may be significantly impaired.

Some examples of triggers for feelings of guilt include limited time with loved ones; infidelity; abuse; and refusing a friend's request.

Mindfulness meditation, which prioritises controlled breathing and deliberate movement, can enhance mind-body awareness and aid in contextualising feelings of guilt. Employing distraction techniques—such as listening to music, reading, or engaging in physical activity—may also prove useful. Addressing guilt proactively through reflection, accepting responsibility, making amends, and implementing positive behavioural changes can facilitate personal development and mitigate the effects of guilt. It is unproductive to persistently focus on previous mistakes; this tendency may be managed by avoiding excessive self-criticism. Furthermore, recognising the unattainability of perfection is important, as pursuing an ideal outcome can lead to unnecessary mental stagnation. Instead, concentrating on achieving the most appropriate solution for your circumstances will enable you to maintain a balanced perspective. There is no simple or immediate remedy for emotional experiences, particularly feelings of guilt. It is beneficial to acknowledge emotions that are justified, rather than attempting to eliminate them. Accepting these emotions can serve as an opportunity for personal growth and guide positive behavioural adjustments in the future.

Chapter Forty-Three

TEACHING YOUR INNER CHILD TO DEAL WITH NARCISSISTIC RELATIONSHIPS

"Somewhere inside, the kid who danced in puddles is still waiting for the rain"

The primary objective of inner child healing is to facilitate the comprehensive processing of both positive and negative early life experiences, thereby supporting individuals in achieving a renewed understanding of their intrinsic value as human beings. Unmet needs during childhood often persist as formative memories that may resurface throughout adulthood, significantly influencing adult identity and behaviour. Regardless of one's background, every individual possesses an inner child requiring care, particularly if unresolved or unprocessed experiences from the past remain. Healing the inner child involves moving forward through rediscovering personal needs, reclaiming lost aspects of oneself, and undertaking a process of self-reparenting to address and mend prior experiences.

Inner child work constitutes the practice of acknowledging, understanding, and addressing wounds originating from childhood. This ongoing process involves the deliberate unlearning of maladaptive behaviours and the adoption of healthier patterns that reinforce constructive coping skills and contemporary self-concepts. It includes reflecting on past experiences and assessing their influence on present-day emotions and actions. By identifying areas of pain and potential growth, individuals can consciously alter habitual responses to emotional triggers provoked by people, situations, or events. This approach emphasises cultivating self-understanding and providing oneself with the emotional support that was needed during childhood but may have been absent. Furthermore, it entails recognising and nurturing vulnerable aspects of oneself with compassion and acceptance. Working with a qualified therapist enables individuals to identify foundational beliefs established during childhood and facilitates the transformation of maladaptive thought patterns into healthier, more compassionate perspectives. This therapeutic process supports healing and empowers individuals with increased resilience and effective strategies for addressing life's challenges. For those with histories of childhood trauma, abuse, or violence, professional intervention is essential in navigating the complexities of recovery, which may require confronting difficult and distressing memories. A central component of this work involves ensuring that the inner child feels secure, protected, and valued—thereby reducing the risk of unresolved issues emerging unexpectedly. Therapeutic approaches often encourage individuals to attune to their inner child, allowing for the acknowledgment and validation of underlying emotions. While many people learn to suppress these feelings during childhood, recognising them in adulthood is crucial. For example, receiving critical feedback may evoke feelings of shame rooted in earlier experiences; by pausing to consider this response, individuals can better understand its origins and approach feedback constructively, emphasising growth rather than self-worth. By identifying triggers for emotional reactions, individuals can develop more effective coping mechanisms; recognising patterns such as a tendency toward abandonment due to formative experiences allows for greater awareness within relationships and improved emotional regulation.

Documenting emotions through journaling and noting typical reactions when the inner child surfaces can reveal behavioural trends. Complementary techniques such as mirror exercises— standing before a mirror and verbalising positive affirmations—

support improvements in self-perception and reinforce self-worth. Another beneficial practice involves writing letters to the inner child, which fosters self-compassion, empathy, and reassurance, thus promoting a sense of safety and trust. Additionally, writing from the perspective of the inner child can highlight unresolved emotions and clarify their impact on daily life. When confronted with triggers related to the inner child, composing supportive statements (for example, 'Dear little me...') can be helpful in validating emotions and reinforcing personal worth and capability. Some individuals find value in meditating with an imagined younger version of themselves, offering affirming messages that would have been meaningful in childhood; this reflective activity enhances identification of factors contributing to emotional difficulties.

It is important to acknowledge both positive and negative dimensions of the inner child, as childhood often encompasses experiences of happiness, security, and playfulness that contribute to worldview formation. Recognising past positive relationships with adults or peers can assist individuals in appreciating and fostering healthy connections in adulthood. Periodically engaging in activities reminiscent of childhood—such as playful behaviours, exploring new interests, or resuming past hobbies—can facilitate personal development and self-reparenting. Individuals who were unable to participate in joyful experiences during childhood might feel disconnected from these aspects later in life; re-engaging with such activities as adults can reinforce the legitimacy of pursuing enjoyment and fulfilment. These steps are designed to enhance awareness, supporting modifications in cognitive patterns and emotional responses. Once insight into the inner child is achieved, this awareness can be used to adjust behaviours and reactions in various contexts. Employing alternative strategies in place of habitual responses may prove advantageous. Clearly communicating needs with a partner—for instance, articulating the desire for physical comfort during challenging moments—can effectively address the requirements associated with one's inner child and promote personal wellbeing.

A critical aspect of this process involves evaluating and challenging outdated belief systems, identifying areas for improvement, and fostering new frameworks grounded in empathy, kindness, and self-compassion. Healing the inner child is a highly individual journey, but its central aim is to enhance awareness of the origins of emotional triggers. Through this understanding, individuals can reduce the impact of these triggers and respond in ways that are consistent with their core values. For example, consider an adult who experienced frequent parental conflict as a child: The coping strategy at that time might have involved withdrawing or attempting to become invisible to avoid exacerbating tension. These adaptive responses often persist into adulthood, manifesting as avoidance of conflict or reluctance to express needs in personal or professional settings. Recognising and changing these patterns— such as asserting boundaries and communicating feelings—signals progress in healing. Another scenario may involve interpersonal relationships in adulthood: If a partner does not reply to a text message, it may evoke feelings of abandonment stemming from earlier life experiences. The initial reaction may be anger, passive-aggressive communication, or catastrophic thinking. However, practicing self-soothing techniques and reframing these thoughts in a constructive way supports emotional regulation and continued personal development.

Indicators of progress in inner child healing may include the ability to respond to situations with greater composure than previously demonstrated, as well as the initiation of self-soothing techniques during interpersonal conflicts. Effectively utilising coping strategies in challenging circumstances can facilitate increased clarity, enabling individuals to express personal needs and emotions—such as grief, guilt, or shame—rather than internalising them. The constructive recognition and management of emotional triggers can lead to observable behavioural improvements. Inner child healing also entails a reassessment of personal values and morals, alongside an evaluation of current requirements for safety, love, and support. This process involves identifying unmet needs and implementing suitable strategies to address them.

Chapter Forty-Four

SELF-COMPASSION

"Unlike self-criticism, which asks if you're good enough, self-compassion asks what's good for you"
Kristin Neff

Individuals who have experienced narcissistic relationships may demonstrate tendencies toward self-criticism and often encounter difficulties with engaging in positive self-talk. The practice of self-compassion is beneficial in addressing mental health concerns, as differentiating between self-critical thoughts and self-kindness can facilitate constructive progress. Understanding self-compassion requires an appreciation of compassion more broadly, which involves a range of feelings, thoughts, motives, desires, urges, and behaviours directed toward the wellbeing of all living beings. Self-compassion is frequently equated with self-love. When individuals habitually prioritise the needs of others at the expense of their own wellbeing, they may experience a depletion of energy and time, consequently reducing their capacity for self-compassion. Such patterns may contribute to a gradual erosion of self-perception, potentially resulting in internal conflict. It is common for individuals in these situations to provide empathy and support to others while neglecting their own needs, leading to ongoing feelings of guilt and unresolved emotional challenges from previous experiences. Terminating a relationship characterised by narcissistic traits should not be viewed as punitive; rather, it represents a necessary step towards supporting one's overall wellbeing. Self-compassion involves approaching oneself with kindness, understanding, and acceptance, akin to the support extended to a respected colleague or friend. By cultivating self-compassion, individuals are likely to experience a reduction in negative self-talk and an improvement in emotional resilience and overall wellbeing.

Oftentimes individuals may engage in self-judgement or self-criticism without clear justification. Acquiring strategies to address these patterns can help in responding to the self with forgiveness, acceptance, and compassion, particularly in difficult situations. Self-kindness refers to consistently treating oneself with patience and understanding, especially when personal expectations are not fulfilled. This attribute can be demonstrated through both actions and thoughts. Examples of self-compassion include offering appropriate support during difficult periods; showing patience and understanding regarding perceived shortcomings; and acknowledging areas for improvement. Recognising common humanity involves viewing oneself as part of a broader human community and understanding that social connection is a feature of human nature. Placing individual experiences within the wider context of the human experience, rather than viewing oneself as isolated, is significant. Examples include accepting that imperfection is a universal condition, acknowledging that everyone has limitations, recognising that emotional responses such as hurt are natural, and understanding that isolation from others is not beneficial.

Understanding that challenges and shortcomings are inherent aspects of the human condition can contribute to increased self-awareness. In this context, mindfulness entails neither avoidance nor excessive identification with one's thoughts and emotions, but rather the capacity to acknowledge them without being compelled to react. Practising self-compassion enables individuals to recognise negative thoughts and feelings without focusing excessively on them, supporting a balance between over-identification and avoidance of discomfort. For instance, this may involve maintaining emotional stability during distressing experiences, remaining objective following setbacks, and approaching emotional states with curiosity during periods of sadness. Changing habitual responses to negative emotions or criticism directed inward often requires ongoing practice. Considering how one would support a friend may offer useful perspective when seeking to apply self-compassion. Validating and acknowledging pain is often used as an initial approach to addressing difficult situations. Identifying mistakes as a routine part of life and responding with understanding, instead of negative self-judgement, can help individuals manage their reactions. Viewing imperfection as a human characteristic may support balanced perspectives on personal flaws and facilitate emotion regulation. Extending similar care to oneself as one does to others may foster empathy and self-understanding. Self-care behaviours can result in physiological effects, including the release of oxytocin. The use of compassionate self-talk or reframing by replacing *I'm such a horrible person for getting upset* with *It's okay that I feel upset*, can promote self-kindness. Recognising both strengths and areas for improvement contributes to self-acceptance, while avoiding exaggeration of faults helps maintain perspective by reinforcing that thoughts and emotions reflect states and behaviours, not permanent personal traits.

Mindfulness practices provide strategies for enhancing focus and awareness. As an element of self-compassion, mindfulness may be developed through activities such as yoga and deep breathing, which are widely accessible in different environments. Avoiding immediate self-judgement and not assuming that behaviours or responses are unchangeable can support changes in thinking patterns. For example, considering the thought *I get grumpy and antisocial while travelling* demonstrates a mindset that links behaviours with certain situations. Reframing these thoughts, such as with *I will do my best to remain calm and to not isolate myself*, may encourage alternative perspectives and support the development of self-compassion. Acknowledging our interconnectedness within the wider context of humanity allows for an appropriate adjustment in perspective. By reducing dependence on external validation, individuals can cultivate a constructive mindset and recognise how perceptions of others' opinions may affect their own thought processes. Avoiding self-critical behaviours—such as expressing frustration about dietary choices or dissatisfaction with personal appearance due to societal pressures—is beneficial. When contentment is not contingent upon external factors, feelings of isolation may be alleviated, thereby supporting the development of social networks vital to overall wellbeing.

Chapter Forty-Five

EMOTIONAL REGULATION

"The calmer you are, the clear you think"

Emotion regulation refers to an individual's capacity to manage and control their emotional state. This may entail strategies such as reframing a challenging situation to minimise feelings of anger or anxiety, concealing outward expressions of sadness or fear, or focusing on factors that promote happiness or calm. Numerous approaches can enhance emotional wellbeing. However, emotion regulation frequently involves *down-regulation*, which aims to decrease the intensity of emotions. For instance, a person experiencing grief may down-regulate their sadness by recalling humorous memories, while someone feeling anxious might distract themselves from distressing thoughts. Additionally, emotion regulation encompasses *up-regulation*, which involves increasing emotional intensity when circumstances—such as imminent danger or obstacles—require heightened levels of anxiety or excitement. Emotion regulation falls into two main categories:

Reappraisal: Modifying one's perspective regarding an emotionally charged event to facilitate a controlled and appropriate response.

Suppression: Suppression is associated with increased negative outcomes and occurs when emotions and thoughts are deeply internalised without efforts to transform negativity into positive results.

Adjusting situations to influence emotional experiences through shifting attention, as well as accepting emotions and reactions, are skills that develop with practice. Establishing routines such as acknowledging emotions, redirecting focus from sources of negative emotions, or reframing situations can contribute to emotional regulation. Rather than avoiding difficult emotions, validating them may create opportunities for learning from setbacks or errors. Consulting with a qualified therapist can assist in developing emotional regulation strategies. Emotions are not always entirely within conscious control, which depends on the aspect of emotional experience being addressed. For example, an individual may initially experience a negative emotion but later reassess its cause or choose to accept it, which may reduce further distress or unwanted behaviours. Alexithymia (emotion control) refers to challenges in identifying and expressing emotions. This characteristic differs among individuals and is associated with psychological distress and difficulties. The regulation of emotions is a fundamental component of effective adult functioning, particularly regarding the management of anxiety and anger in accordance with societal norms. Inadequate emotional control may lead to behaviours or remarks that individuals subsequently find regrettable, underscoring the necessity for proficient emotion regulation strategies. Persistent emotional dysregulation is evident in various mental health disorders and can significantly impact personal wellbeing as well as interpersonal relationships.

Anger, resentment, and disappointment are prevalent emotional states commonly encountered. While these emotions are manageable, they are not necessarily considered pathological in all cases. However, persistent challenges in regulating emotions may signal an underlying mental health disorder, such as borderline personality disorder or depression. Poor regulation of emotions including anger, anxiety, or fear can adversely affect interpersonal relationships by causing overreactions, distress, or missed opportunities. Furthermore, research indicates that consistently suppressing emotions is associated with reduced wellbeing and lower relationship satisfaction. Multiple factors influence emotion regulation, including individual's beliefs regarding negative emotions, the accessibility of emotion-regulation skills, and the presence of situations that provoke strong affective responses. For instance, situation selection represents one strategy within emotion regulation. Circumstances marked by uncertainty or perceived threats may heighten the risk of emotional flooding and diminish self-regulatory capacity.

Chapter Forty-Six

MONITORING EMOTIONS: EMPATHY'S EVIL TWIN

"Emotion can be the enemy, if you give into your emotion, you lose yourself. You must be at one with your emotions, because the body always follows the mind"
Bruce Lee

Empathy, the ability to understand and share another person's point of view is an important human competency. It strengthens interpersonal relationships, encourages cohesion, and contributes to a greater sense of inclusion. Referred to as theory of mind, this faculty is essential for maintaining robust mental health and overall wellbeing. In comparison, emotion monitoring pertains to the ongoing assessment of others' emotional states and represents a distinct yet related concept. Unlike empathy, which arises in response to specific external cues related to another individual's situation, emotion monitoring involves continual—often unconscious—observation aimed at recognising potential negative emotions in others. The intent is to anticipate discomfort for these individuals. In some interpersonal situations, such as those involving narcissistic people, one partner's tendency to monitor emotions may be used, by the other, for their own purposes. Over time, this anticipation can lead to reduced focus on one's own emotional state.

Those who have experienced narcissistic abuse may consistently monitor the emotional states of those around them. This behavioural pattern is frequently noted among women raised in environments where other people's needs were prioritised above their own. For example, someone who grew up with a parent susceptible to anger may become acutely attuned to that parent's moods, often suppressing personal emotions to prevent conflict. Over time, this heightened focus on other's emotions can develop into a broader tendency toward emotion monitoring across various relationships. In adulthood, such individuals may struggle to recognise their own needs or engage authentically, as their attention remains primarily directed toward managing the emotions and responses of others. A habitually maintained heightened awareness of the emotions of others may result in a person losing touch with their own internal experiences: Attunement to one's personal emotional landscape becomes compromised in favour of attending to the needs of others, which can negatively impact self-concept, self-esteem, and overall mood. Focusing primarily on external emotional cues increases the risk of neglecting personal needs, leaving individuals without reciprocal emotional support. As a result, persistently monitoring the feelings of others can become highly exhausting and emotionally depleting. Such tendencies often arise as adaptive responses to challenging environments to ensure safety. Even after these circumstances improve, the propensity to monitor other's emotions can persist. When individuals become conscious of their emotion monitoring habits and start to balance their focus, this transition represents an important development in self-awareness.

Reducing excessive emotional monitoring may influence spontaneity in social interactions. This method can affect authenticity, allowing individuals to respond to others according to personal feelings and reciprocate as needed. Limiting unnecessary monitoring may result in less demanding interpersonal exchanges and could alter motivation for social engagement. These factors might impact self-esteem and symptoms linked to depression and social anxiety.

Chapter Forty-Seven

SMALL STEPS TOWARD MOVING ON

---

*"Revenge is like a rolling stone, which, when a
man hath forced up hill, will return upon him
with a greater violence, and break those bones
where sinews gave it motion"*
Jeremy Taylor

The process of healing is often neither straightforward nor seamless. Individuals may encounter recurring emotions they previously believed were resolved, causing them to devote substantial time to contemplating the decisions required for establishing healthy boundaries. It is important to recognise that repeating certain behaviours while adopting new strategies, as well as alternating between periods of progress and emotional challenge, are common aspects of this journey. Even when individuals feel assured of their decisions, it is typical to re-evaluate those choices periodically. Accepting responsibility for one's mistakes with self-compassion and pursuing personal growth with understanding are essential components of the healing process. It is important to note harmful patterns of behaviour in relationships: Constant criticism; draining the victim's energy; always playing the victim; never apologising; manipulating others' emotions; dismissing the feelings of others; creating unnecessary drama; competing and never supporting; guilt tripping; blaming others, always; disrespecting boundaries; spreading negativity constantly; ignoring the achievements of others;

talking behind backs; gaslighting reality; creating anxiety; giving fake kind gestures; showing no empathy; taking, never giving; are jealous of success; lying without remorse; laughing at pain; twisting words to suit them; controlling choices; creating doubt; sabotaging growth; instigating unnecessary conflict; making it about themselves; refusing to listen; and celebrating perceived failures of others.

Engagement in relationships with individuals exhibiting narcissistic characteristics may lead to substantial emotional and psychological consequences. Identifying manipulative behaviours and relinquishing unrealistic expectations are critical components of the recovery process. Permitting oneself to acknowledge and process previous losses facilitates the cultivation of sound judgment and supports healing. Those with narcissistic tendencies frequently demonstrate reduced empathy and a consistent focus on their personal needs, occasionally to the detriment of others. These priorities, often concealed through manipulative strategies such as gaslighting and projection, often aim to influence or control relationship dynamics. Following the conclusion of such relationships, these individuals might experience or express perceptions of betrayal or abandonment, further exacerbating confusion and emotional distress for former partners. These patterns may result in enduring effects, including diminished self-esteem and increased anxiety during future relationship formation. The following are key factors in the impact of narcissistic relationships:

Demystification: Identifying patterns of control and abuse is a critical component in understanding personal experiences and reestablishing self-awareness. Individuals exhibiting narcissistic traits may demonstrate consistent behaviours, such as initiating relationships with romantic gestures followed by diminished support or respect. These indicators are often more discernible after the relationship concludes. Constructing a relationship timeline can provide valuable insights, detailing positive beginnings, the onset of behavioural changes, and alterations in priorities and responses to expressed concerns.

Documenting any uncertainties and unresolved issues can also be advantageous. Sharing this information with a trusted confidant or a mental health professional is recommended.

Disillusionment: Certain individuals may emphasise positive attributes in others. This can, occasionally, be leveraged by those with narcissistic characteristics. Failure to recognise these interpersonal dynamics may lead to increased vulnerability to manipulation. Individuals demonstrating narcissistic tendencies often seek control rather than fostering mutual respect. It is essential to acknowledge if a partner's behaviours have deviated from your expectations or values as part of the personal growth process. Compiling a comprehensive list of challenges or unhealthy aspects within the relationship can facilitate greater insight into both the partner and the relationship itself. Recording specific instances of discomfort, unhelpful patterns, and personal compromises contributes to establishing a more objective perspective when reviewing past experiences.

Decoupling: Coping with the end of a relationship involves various strategies, such as practicing mindfulness and self-awareness. Maintaining distance from an ex-partner can assist in adjusting to the change. Identifying and listing emotions associated with the breakup can clarify individual responses, which may include sadness, guilt, anger, shame, hurt, longing, loneliness, or concerns about future relationships. Emotions like guilt, anger, shame, and hurt may indicate ongoing thoughts about the previous relationship—feelings that are typical during the adjustment process. Longing and loneliness may reflect interest in future social connections. Acknowledging and processing these emotions can contribute to personal adaptation. This goes a long way in helping negative memories related to the relationship becoming less prominent over time.

Discernment: Developing an understanding of the unhealthy dynamics present in previous relationships enables individuals to identify potential concerns or desirable qualities in future partnerships. This self-awareness facilitates the recognition of recurring patterns, both negative and positive, and provides a basis for referencing specific behaviours that were not conducive to a healthy relationship. Positive attributes can often be identified as the opposite of negative indicators; for example, if a prior partner was not attentive, seeking someone who values effective communication and emotional respect may prove advantageous. Likewise, past experiences with one-sided interactions may underscore the importance of mutual consideration. Evaluating observable actions rather than relying solely on verbal assurances supports more informed and balanced decision-making.

Healing And Revitalisation: The final phase of healing necessitates shifting focus from a previous partner and relationship history toward personal growth and future objectives. It is beneficial to reflect on how past experiences have shaped trust in oneself and others, as well as to assess their influence on perceptions of love and romantic partnerships. Additionally, reevaluating any changes in self-perception may provide valuable insight. Adopting more constructive perspectives in place of limiting assumptions promotes a thorough reassessment of views regarding love and relationships. Defining one's identity within the context of romantic connections—including goals, needs, boundaries, discernment, and self-respect—can facilitate continued progress. Prioritising future aspirations over past events fosters meaningful advancement.

Chapter Forty-Nine

HEALING FROM NARCISSISTIC PARENTS

"A child won't say 'get over it' when an adult cries. There will be no punishment, or judgement from them. Instead, they might hug you, stay by your side or lend you their favourite toy. They deserve the same treatment" Donna Siggers

---

Having one parent who exhibits abusive behaviour and another who avoids confrontation or enables the situation means both may contribute to the overall impact on the child. While a non-confrontational parent may act out of self-protection, their lack of intervention can result in emotional effects such as feelings of abandonment. These circumstances can influence trust and relationships later in life. Passive parents may not always recognise their involvement, which can affect the healing process. Both the abusing and non-intervening parents play roles in shaping a child's experiences, and the need for personal safety is understandable. Research indicates that children exposed to domestic abuse may experience brain changes comparable to those seen in combat veterans. Prolonged stress caused by unresolved anger may negatively influence physical health, including sleep quality, heart function, cholesterol levels, and blood pressure. Individuals affected by being raised by a parent with narcissistic traits may seek recovery through forgiveness, which can contribute to reducing anger, lowering anxiety, alleviating depression, improving self-esteem, fostering hope for the future, and managing stress. These factors are commonly recognised as beneficial for mental health and overall wellbeing.

Children of parents diagnosed with narcissistic personality disorder, or those exhibiting narcissistic traits, may be affected by intergenerational trauma patterns. In some cases, these parental characteristics are associated with adverse experiences during the parents' own childhoods, including factors such as childhood neglect or various forms of abuse, such as verbal, physical, sexual, and/or emotional abuse. Household dysfunction can influence behavioural development. Narcissistic patterns transmitted across generations—a phenomenon referred to as intergenerational trauma—often result in experiences of grief and loss. Society may not always recognise the emotions associated with intergenerational trauma, categorising them as forms of disenfranchised grief. Addressing and processing these emotions is a crucial aspect of recovery. Additionally, some individuals may self-disenfranchise by failing to acknowledge the impact of losing a childhood uninfluenced by narcissistic dynamics. Features that are potentially missing in childhood include a lack of: Nurturing parenting that was deserved; carefree and innocent childhood experienced by peers; a stable environment enabling a sense of security; unconditional acceptance to foster healthy self-esteem; and freedom to develop self-identity. Acknowledging and addressing significant losses is a critical step in advancing the healing process. Acquiring an understanding of narcissism enables individuals to identify detrimental behaviours, including gaslighting and emotional abuse. The tendency to internalise criticism or assume responsibility for a parent's dissatisfaction can adversely affect self-esteem. A comprehensive knowledge of this condition facilitates the development of effective strategies for navigating interactions involving narcissistic behaviours.

Establishing boundaries constitutes an essential aspect of self-protection. Internal boundaries serve to mitigate the influence of parental behaviours on individual self-perception by reframing unkind actions as indicative of narcissism rather than personal deficiencies, thereby reducing the likelihood of internalising adverse effects. External boundaries involve moderating excessive accommodation toward a parent, which may include respectfully declining requests and articulating one's own preferences and opinions. It is noteworthy that setting external boundaries often becomes more feasible in adulthood, particularly when one no longer shares a household with the parent concerned. Demonstrating autonomy in decision-making—such as changing jobs or selecting academic programmes according to personal interests—reflects the capacity for self-directed choice. While some individuals may initially pursue avenues aligned with parental aspirations, these may not necessarily correspond with their own goals. Pursuing an alternative path supports the development of personal identity, alignment with individual objectives, and the establishment of boundaries from potentially manipulative influences.

For those whose parents exhibit narcissistic traits, discerning the elements of healthy relationships can be challenging. Acquiring a comprehensive understanding of narcissism and its impacts can facilitate the ability to identify positive social interactions. Applying this insight assists in forming constructive friendships and romantic partnerships, which is integral to modifying established behavioural patterns. Engaging a therapist with expertise in narcissistic relationship dynamics may further support the development of effective coping strategies. They can assist by identifying narcissistic behaviours, changing thought patters, setting boundaries, and identifying healthy relationship characteristics.

While many parents strive to do their best, parenting does not come with explicit guidelines, and necessary skills are frequently developed over time. This, however, does not diminish the complexities or difficulties that may arise from certain parenting approaches. Recognising that caregivers may have demonstrated care or made sacrifices while simultaneously causing harm or lacking the emotional maturity required to meet specific needs can be challenging. Both beneficial and adverse effects of caregiving often coexist and continue to shape adult relationships. Exposure to environments characterised by chaos, criticism, emotional distance, or unavailability can influence the development of an individual's nervous system. These circumstances may also foster enduring beliefs that continue to shape individual's perspectives regarding their value, worth, and the significance of their needs, desires, and boundaries. Recognising past experiences is intended to facilitate understanding rather than to assign blame; it involves reflecting on and discussing these events in a constructive manner. The emphasis is not on rejecting others but on fostering personal recovery. It is important to balance acknowledging past difficulties with moving forward productively. Understanding relational dynamics requires recognising that certain outcomes were not the individual's fault. Yet addressing them remains a personal responsibility. Forgiveness has the potential to mitigate negative effects without necessitating approval of inappropriate actions, allowing individuals to acknowledge harm while choosing to forgive.

Chapter Fifty

ROMANTIC RELATIONSHIPS FOLLOWING CHILDHOOD SEXUAL ABUSE

"If you do not address your childhood traumas,
your romantic relationships will"
Neil Strauss

Individuals who have survived childhood abuse frequently face complex emotional and psychological difficulties upon reaching adulthood. In addition to addressing the original trauma, these individuals may experience enduring effects that manifest over time. The repercussions of childhood sexual abuse are particularly intricate; when the perpetrator is known to the survivor, substantial feelings of betrayal can arise. Survivors often describe a sense of powerlessness due to their inability to prevent the abuse, along with societal stigma related to victimisation. Sexual trauma may present as either heightened sexual activity or dysfunction. Such experiences can profoundly impact survivor's perspectives, resulting in challenges with trust—both self-trust and trust in others. Such challenges can affect survivor's ability to form healthy, committed relationships in adulthood. Individuals with a history of childhood sexual abuse may view adversity in life and relationships as overwhelming, which may increase vulnerability to patterns of negative thoughts and behaviours. Negative perceptions of self-worth and trust in others are common, potentially perpetuating maladaptive responses if not addressed. Survivors might also exhibit limited self-protective skills, sometimes maintaining a victim identity rather than adopting a resilient survivor perspective, which can continue to impact their vulnerability to exploitation. The lasting effects of early sexual abuse often result in the absence of secure and supportive childhood experiences, potentially hindering the development of adaptive coping mechanisms and positive outcomes in adulthood.

Adult survivors of childhood sexual abuse may encounter difficulties in managing interpersonal and romantic relationships, which are often more complex than other relational forms. Foundational familial relationships, when marred by mistrust, can pose challenges for survivors seeking to establish trust with others. Intimacy following childhood sexual abuse can impact sexual desire, arousal, and orgasm, as these experiences may be associated with prior negative events. Empirical research demonstrates that survivors often display distinct patterns of engagement in intimate relationships, including an increased prevalence of risky sexual behaviours such as multiple partnerships, unprotected sex, unplanned pregnancies, and higher rates of sexually transmitted diseases. The long-term stability of relationships can be compromised if the effects of past abuse remain unaddressed. Survivors may report heightened feelings of isolation and lower relationship satisfaction compared to individuals without a history of abuse. Relationship dynamics reminiscent of those present during the period of abuse may trigger emotional responses within current partnerships, contributing to ongoing interactional cycles that affect both individual's sense of control, connection, and empowerment. For some, intimate encounters may also provoke distressing memories. A sustained belief that intimacy is inherently risky and that others are untrustworthy can further challenge the development of secure attachments. Survivors might also experience struggles with self-worth and perceptions of being undeserving of love, factors which can undermine relational stability throughout their lives. These struggles include feelings of unworthiness, feeling dirty, undesirable, depression, self-doubt, shame, PTSD symptoms, anorgasmia, dissociation during sex, distrust of partner's motives, intense emotions, somatic abuse memories, unconscious abuse-driven actions, difficulty expressing emotions, struggling to accept love, and avoidant coping. Children process emotional experiences in a manner comparable to adults. The formation of their identities is shaped by internalising and reflecting upon the attitudes, behaviours, and expectations within their environment, with sensitivity to adverse circumstances and negative influences from significant role-models or caregivers.

Therapeutic interventions are instrumental in empowering survivors to reestablish autonomy, regulate emotional reactions to triggers, and improve the overall quality of romantic relationships. Trauma-informed therapy is particularly effective for couples, as it facilitates an understanding of how previous abuse or neglect may influence current relational dynamics. This modality allows clinicians to provide targeted guidance that assists individuals and couples in distinguishing between past experiences and present circumstances. Optimal progress is normally achieved through a combination of individual and joint therapeutic sessions. By engaging in trauma-informed practice, partners develop enhanced skills for understanding each other's perspectives, recognising the persistent effects of trauma, and implementing healthier methods for processing thoughts and emotions within their relationship.

Chapter Fifty-One

ASSERTIVENESS

"Being assertive does not mean attacking or ignoring other's feelings. It means that you are willing to hold up for yourself fairly—without attacking others"
Albert Ellis

Experiencing challenges in expressing opinions, encountering others who may dismiss or undermine one's views, lacking confidence when speaking, or handling situations with aggression are all examples where developing assertiveness skills can be beneficial. Assertiveness involves standing up for oneself in a clear, calm, and confident manner. It is a skill that assists individuals in managing themselves, others, and various situations. It can be useful for influencing others to achieve acceptance, agreement, or behavioural changes. Assertive communication allows people to express opinions in a clear and confident manner. Those who practice assertiveness are direct with themselves and those around them. There is a distinction between assertiveness and aggression, though the two are sometimes confused. Assertiveness involves finding a balance—being straightforward about one's needs and wants while also considering the perspectives of others. With assertiveness, individuals communicate their points directly and respectfully. There are numerous advantages to adopting assertive behaviour. Assertiveness enables individuals to communicate their needs and preferences clearly and authoritatively, while maintaining fairness and empathy. It also contributes to greater self-confidence and supports positive mental health. Assertive individuals often excel as managers, as they treat others with respect and equity, fostering reciprocal relationships and earning them reputations as approachable and effective leaders. Their capacity to negotiate mutually beneficial outcomes arises from recognising and valuing differing perspectives, allowing them to find common ground. Furthermore, assertive people are adept at problem-solving, empowered to pursue the most appropriate solutions to the challenges they encounter. In general, those who demonstrate assertiveness tend to experience lower levels of anxiety or stress, possess a strong sense of self-assurance, and are less likely to feel threatened or victimised when circumstances do not unfold as anticipated. They report higher job satisfaction and can effectively maintain boundaries by discerning when to agree with a person while declining a specific request.

However, excessive assertiveness can be detrimental. Disregarding input from others can alienate colleagues and undermine personal relationships. Practising assertiveness incrementally within professional or personal environments can help strengthen one's skills and self-esteem, clarifying the appropriate moments to employ assertive communication. While developing assertiveness can present challenges, it remains an attainable skill. Developing a clear understanding of oneself, coupled with a strong belief in one's intrinsic value and contributions to relationships is fundamental for assertiveness. This self-belief forms the foundation of both self-confidence and assertive conduct, empowering individuals to recognise their right to dignity and respect. Such confidence enables people to advocate for their rights, set appropriate boundaries, and uphold their personal values, goals, and needs. It is essential, however, for self-confidence to remain distinct from self-importance. A person's rights, thoughts, feelings, needs, and desires hold equal significance alongside those of others; no individual should be considered as superior. Expressing needs and wants clearly can help build assertiveness, enabling individuals to reach their potential. This involves recognising and addressing personal priorities. Setting clear goals is one way to work towards achieving intended outcomes and exceeding initial expectations. Success can result from maintaining sustained attention and commitment to tasks. Confidence in one's assertiveness is advantageous in the workplace, facilitating more effective interactions and advocacy for personal needs and interests. When needs cannot be immediately met, supervisors should demonstrate both the ability and willingness to support long-term objectives. This strategy promotes sustained progress toward future goals. It is important to advocate your endeavours confidently without forcing sacrifice from the need of others. These business-place techniques are transferable skills that can be used in intimate relationships.

Recognising the importance of individuals' behaviours are, ideally, not controlled by external influences. A neutral approach involves acknowledging other's perspectives, even when they differ from one's own. Assertiveness shouldn't be misconstrued as reciprocating anger; rather, it entails remaining composed and showing respect during challenging interactions, whilst ensuring that the needs of all parties are considered. Individuals should have both the ability and autonomy to express themselves appropriately, addressing difficult subjects constructively and with sensitivity to achieve positive outcomes. It is acceptable for all to assert their position and respond to individuals who question the rights of all involved. Experiencing anger is natural, but it is important to regulate emotions and maintain respect throughout interactions.

Demonstrating openness to criticism and receiving both positive and negative feedback with humility and composure is essential for personal development. It is important to foster an environment where open discussion can take place without resorting to defensiveness or anger. This approach to feedback allows individuals to look beyond emotional responses and leverage constructive criticism for meaningful improvement. Similarly, developing the ability to assertively decline requests, especially when it is not customary, supports greater self-assurance. After each attempt at assertiveness, take a few minutes to reflect by considering questions such as, 'how did I manage that situation', 'what did I do effectively' and 'what adjustments could I make in the future'. This practice will support ongoing development and enable the identification of specific areas for improvement. Encountering setbacks should be used as a learning opportunity rather than a discouragement. Recognising achievements while maintaining perspective on challenges is essential for sustained growth.

It is essential to acknowledge and consider the other party's perspective regarding the situation. After their viewpoint has been considered, communicating needs clearly and respectfully by employing statements such as 'I want,' 'I need,' or 'I feel' can help convey position and feeling in a direct and assertive manner. This could include reiterating concerns more firmly—while remaining courteous and respectful—or specifying subsequent measures. If reasonable requests are not fulfilled, delaying a response can provide time to consider alternative approaches and generate a more measured reply. Preparing thoughts ahead of time may contribute to more effective solutions than responding without reflection. Using definite and emphatic verbs in communication can clarify the intended message and reduce ambiguity. For example, using 'will' rather than 'could' or 'should'; 'want' instead of 'need'; or 'choose to' in place of 'have to', can enhance precision. It is important to ensure that persistence does not become repetitive or forceful communication. While persistence can help prevent exploitation, it may also be interpreted as aggressive behaviour, which could result in others perceiving manipulation. Formulating a mental script enables clear and confident communication of one's feelings. This strategy facilitates advance preparation by employing a four-step framework to articulate:

The event: Clear communication of perspective regarding the situation or issue to the other party.

Feelings: Describe feelings and express emotions clearly.

Needs: Clearly communicating expectations helps prevent misunderstandings.

The consequences: Clearly articulate the beneficial outcomes of the request.

Assertiveness involves striking an appropriate balance between passivity, characterised by sufficient self-advocacy, and aggression, which manifests as hostile or domineering behaviour. Being assertive demonstrates a clear sense of self-worth and the recognition that one's needs and objectives are valid. It entails standing up for oneself even in challenging circumstances, without resorting to dominating or disregarding others. Acting solely in one's own interest, at the expense of the rights, feelings, or needs of others, constitutes aggression rather than assertiveness. Developing assertiveness is an attainable skill that can be cultivated over time through recognising personal needs, expressing them constructively, and learning to set boundaries when necessary. Employing assertive communication techniques can support individuals in articulating their thoughts and emotions clearly, confidently, and respectfully. Change does not occur instantly; however, by consistently applying these techniques, individuals can gradually develop the confidence and self-assurance necessary for assertiveness. Additionally, this process may lead to increased productivity, efficiency, and respect from others.

Chapter Fifty-Two

END NOTE

We've learnt that narcissistic traits can cause unpredictable behaviours, often prompting victims to withdraw socially and that, despite variations in timing and presentation, these behaviours typically follow discernible patterns. Recognising these patterns can yield valuable insights and promote a deeper understanding of interpersonal dynamics. While there might be concern that analysing the motivations behind abusive conduct could be misconstrued as justification, it remains important to trust in self-resilience when addressing these challenges. An enhanced comprehension of abusive patterns can facilitate healthy emotional detachment from the abuser as well as personal growth and resilience and, thereby, supporting self-esteem, wellbeing, and self-respect.

A significant obstacle in understanding narcissistic behaviour is overcoming initial discomfort and apprehension toward the process. This requires faith in the potential benefits of increased recognition of the impact it has. Subsequently, individuals must set aside personal emotions, reactions, and protective barriers to allow for objective reflection. Defensive responses can impede progress; therefore, suppressing internal resistance—such as doubts about the validity or value of the process—is essential. The most mentally demanding aspect often involves applying acquired knowledge to adopt the narcissist's perspective, even briefly. This exercise demands considerable focus and cognitive effort and should be undertaken only for short durations. Attempting to interpret events through another's worldview, especially one as distinct as a narcissist's, necessitates maintaining multiple perspectives simultaneously. It is critical to note that understanding another's reasoning does not equate to agreement with their perspective. Adopting a contrasting belief system momentarily enables one to trace the logic underlying their actions. However, this adoption is strictly temporary and should not result in acceptance of flawed premises. Logical arguments are constructed from foundational assumptions, supporting statements, and subsequent conclusions. Narcissists often operate from questionable assumptions, leading to conclusions that may not align with reality. By examining these assumptions critically, one can follow the reasoning without accepting its validity. The initial step in this process involves increasing awareness of situational dynamics through observation, without attempting to alter them. Next, acquiring knowledge about narcissistic behaviours establishes a foundation for confidence and stability. Applying this understanding enables recognition of behavioural patterns, helping to track progression from baseline to triggering events: Observing these chains of reaction with curiosity rather than judgement fosters objective analysis. Developing the habit of questioning assumptions encourages deeper inquiry into each interaction. By synthesising observations and acquired knowledge, individuals become adept at identifying patterns and resolving conflicts as they arise. With continued practice, shifting perspectives becomes more efficient. One can temporarily view situations through the narcissist's lens to interpret specific behaviours before returning to their own perspective.

While increased understanding does not resolve narcissistic behaviours, it can reduce internal distress and foster a sense of calm conducive to clear decision-making. These elements form the basis of empowerment and personal agency. They enable new pathways for exploration and a more fulfilling life that promotes feelings of safety, mindfulness, and contentment.

ABOUT THE AUTHOR

Donna Siggers qualified as a psychotherapist in 2010 and further expanded her expertise through mature studies in psychology and criminology. Following a brain injury in 2014, that significantly impacted her life, she began writing as part of her recovery process.

Donna's first novel, *Broken* received an international award willing her to continue writing crime drama. With her profound interest in mental health, she also writes non-fiction with the objective of reducing the stigma associated with mental illness while supporting others in overcoming adversity.

BOOKS BY DONNA SIGGERS

MENTAL HEALTH BOOKS

Lost Soul: Broken Soul to Soul on Fire

Soul Searching: To PTSD Hell and Back Twice

Free Spirit: How to Break Beyond Limitations

You're Not Mad: Understanding and Dealing with Narcissism

NOVELS

THE WARWICK COOPER THRILLERS

Broken

Betrayal

Bound

THE BLANDFORD THRILLERS

Faces of the Missing

Printed in Dunstable, United Kingdom